NANOTECHNOLOGY SCIENCE AND TECHNOLOGY

DEVELOPMENT AND APPLICATION OF NANOFIBER MATERIALS

NANOTECHNOLOGY SCIENCE AND TECHNOLOGY

Safe Nanotechnology
Arthur J. Cornwelle)
2009. 978-1-60692-662-8

National Nanotechnology Initiative: Assessment and Recommendations
Jerrod W. Kleike (Editor)
2009. 978-1-60692-727-4

Nanotechnology Research Collection - 2009/2010. DVD edition
James N. Ling (Editor)
2009. 978-1-60741-293-9

Strategic Plan for NIOSH Nanotechnology Research and Guidance
Martin W. Lang
2009. 978-1-60692-678-9

Safe Nanotechnology in the Workplace
Nathan I. Bialor (Editor)
2009. 978-1-60692-679-6

Nanotechnology in the USA: Developments, Policies and Issues
Carl H. Jennings (Editor)
2009. 978-1-60692-800-4

Nanotechnology: Environmental Health and Safety Aspects
Phillip S. Terrazas (Editor)
2009. 978-1-60692-808-0

New Nanotechnology Developments
Armando Barrañón (Editor)
2009. 978-1-60741-028-7

Electrospun Nanofibers and Nanotubes Research Advances
A. K. Haghi (Editor)
2009. 978-1-60741-220-5

Nanotechnology Research Collection - 2009/2010. PDF edition
James N. Ling (Editor)
2009. 978-1-60741-292-2

Carbon Nanotubes: A New Alternative for Electrochemical Sensors
Gustavo A. Rivaș, María D. Rubianes, María L. Pedano, Nancy F. Ferreyra, Guillermina Luque and Silvia A. Miscoria
2009. 978-1-60741-314-1

Polymer Nanocomposites: Advances in Filler Surface Modification Techniques
Vikas Mittal (Editor)
2009. 978-1-60876-125-8

Nanostructured Materials for Electrochemical Biosensors
Yogeswaran Umasankar; S. Ashok Kumar; Shen-Ming Chen (Editors)
2009. 978-1-60741-706-4

Magnetic Properties and Applications of Ferromagnetic Microwires with Amorphous and Nanocrystalline Structure
Arcady Zhukov and Valentina Zhukova
2009. 978-1-60741-770-5

Electrospun Nanofibers Research: Recent Developments
A.K. Haghi (Editor)
2009. 978-1-60741-834-4

Nanofibers: Fabrication, Performance, and Applications
W. N. Chang (Editor)
2009. 978-1-60741-947-1

Barrier Properties of Polymer Clay Nanocomposites
Vikas Mittal
2009. 978-1-60876-021-3

Bio-Inspired Nanomaterials and Nanotechnology
Yong Zhou (Editor)
2009. 978-1-60876-105-0

Nanofibers: Fabrication, Performance, and Applications
W. N. Chang (Editors)
2009. 978-1-61668-288-0

Nanotechnology: Nanofabrication, Patterning and Self Assembly
Charles J. Dixon and Ollin W. Curtines (Editors)
2010. 978-1-60692-162-3

Gold Nanoparticles: Properties, Characterization and Fabrication
P. E. Chow (Editor)
2010. 978-1-61668-009-1

Micro Electro Mechanical Systems (MEMS): Technology, Fabrication Processes and Applications
Britt Ekwall and Mikkel Cronquist (Editors)
2010. 978-1-60876-474-7

Nanomaterials: Properties, Preparation and Processes
Vinicius Cabral and Renan Silva (Editors)
2010. 978-1-60876-627-7

Nanopowders and Nanocoatings: Production, Properties and Applications
V. F. Cotler (Editor)
2010. 978-1-60741-940-2

Nanomaterials Yearbook - 2009 . From Nanostructures, Nanomaterials and Nanotechnologies to Nanoindustry
Gennady E. Zaikov and Vladimir I. Kodolov (Editors)
2010. 978-1-60876-451-8

Nanoparticles: Properties, Classification, Characterization, and Fabrication
Aiden E. Kestell and Gabriel T. DeLorey (Editors)
2010. 978-1-61668-344-3

Nanoporous Materials: Types, Properties and Uses
Samuel B. Jenkins (Editor)
2010. 978-1-61668-182-1

Mechanical and Dynamical Principles of Protein Nanomotors: The Key to Nano-Engineering Applications
A. R. Khataee and H. R. Khataee
2010. 978-1-60876-734-2

Electrospun Nanofibers and Nanotubes Research Advances
A. K. Haghi (Editor)
2010. 978-1-60876-762-5

TiO2 Nanocrystals: Synthesis and Enhanced Functionality
Ji-Guang Li , Xiaodong Li, Xudong Sun
2010. 978-1-60876-838-7

Nanomaterial Research Strategy
Earl B. Purcell (Editor)
2010. 978-1-60876-845-5

Magnetic Pulsed Compaction of Nanosized Powders
G.Sh. Boltachev, K.A. Nagayev, S.N. Paranin, A.V. Spirin and N.B. Volkov
2010. 978-1-60876-856-1

Nanostructured Conducting Polymers and their Nanocomposites: Classification, Properties, Fabrication and Applications
Ufana Riaz and S.M. Ashraf
2010. 978-1-60876-943-8

Phage Display as a Tool for Synthetic Biology
Santina Carnazza and Salvatore Guglielmino
2010. 978-1-60876-987-2

Bioencapsulation in Silica-Based Nanoporous Sol-Gel Glasses
Bouzid Menaa, Farid Menaa, Carla Aiolfi-Guimarães and Olga Sharts
2010. 978-1-60876-989-6

ZnO Nanostructures Deposited by Laser Ablation
M. Martino, D. Valerini, A.P. Caricato,
A. Cretí, M. Lomascolo, R. Rella
2010. 978-1-61668-034-3

Development and Application of Nanofiber Materials
Shou-Cang Shen, Wai-Kiong Ng, Pui-Shan Chow
and Reginald B.H. Tan
2010. 978-1-61668-931-5

Polymers as Natural Composites
Albrecht Dresdner and Hans Gärtner (Editors)
2010. 978-1-61668-168-5

Synthesis and Engineering of Nanostructures by Energetic Ions
Devesh Kumar Avasthi and Jean Claude Pivin (Editors)
2010. 2978-1-61668-209-5

From Gold Nano-Particles Through
Nano-Wire to Gold Nano-Layers
V. Švorčík, Z. Kolská, P. Slepička
and V. Hnatowicz
2010. 978-1-61668-316-0

Gold Nanoparticles: Properties, Characterization and Fabrication
P. E. Chow (Editor)
2010. 978-1-61668-391-7

Nanoporous Materials: Types, Properties and Uses
Samuel B. Jenkins (Editor)
2010. 978-1-61668-650-5

Phase Mixture Models for the Properties of Nanoceramics
Willi Pabst and Eva Gregorova
2010. 978-1-61668-673-4

Applications of Electrospun Nanofiber Membranes for Bioseparations
Todd J. Menkhaus, Lifeng Zhang and Hao
2010. 978-1-60876-782-3

Development and Application of Nanofiber Materials
Shou-Cang Shen, Wai-Kiong Ng, Pui-Shan Chow and Reginald B.H. Tan
2010. 978-1-61668-829-5

Polymers as Natural Composites
Albrecht Dresdner and Hans Gärtner (Editors)
2010. 978-1-61668-886-8

Phase Mixture Models for the Properties of Nanoceramics
Willi Pabst and Eva Gregorova
2010. 978-1-61668-898-1

NANOTECHNOLOGY SCIENCE AND TECHNOLOGY

DEVELOPMENT AND APPLICATION OF NANOFIBER MATERIALS

SHOU-CANG SHEN
WAI-KIONG NG
PUI-SHAN CHOW
AND
REGINALD B.H. TAN

Nova Science Publishers, Inc.
New York

For permission to use material from this book please contact us:
Telephone 631-231-7269; Fax 631-231-8175
Web Site: http://www.novapublishers.com

LIBRARY OF CONGRESS CATALOGING-IN-PUBLICATION DATA

Development and application of nanofiber materials / Shou-Cang Shen ... [et al.].

p. cm.

Includes index.

ISBN 978-1-61668-931-5 (softcover)

1. Nanofibers. I. Shen, Shou-Cang.

TA418.9.F5D48 2010

620'.5--dc22

2010012161

Published by Nova Science Publishers, Inc. + New York

CONTENTS

Preface		**xiii**
Chapter 1	Preparation Methods of Nanofibers	**1**
Chapter 2	Properties and Applications of Nanofiber Materials	**51**
Chapter 3	Risk Assessment	**79**
Chapter 4	Summary	**83**
References		**85**
Index		**131**

PREFACE

One-dimensional structured (1D) nanofiber materials have attracted great interest over the past decade because they can potentially address many advanced applications involving dimensionality and size-confined quantum phenomena. Compared with non-dimensional nanoparticles, the 1D structure of nanofiber materials lends them the anisotropic properties suitable for fabrication of nanodevices. They are expected to play a crucial role as interconnects and functional units in fabrication of lab-in-chip, electronics, electronic-chemical, optoelectronic and electromechanical devices with a nanoscale dimension. In addition, nanofiber materials have been found to have broad application potentials in various other areas, such as advanced ceramics, metal/composite reinforcement, membrane separation, sensor, catalysis, biological and biomedical fields.

1D nanomaterials with different aspect ratios and morphologies can be generally classified as nanorods, nanowires, and nanotubes. Nanorods are usually straight 1D nanostructured materials with a diameter of 1~100 nm and low aspect ratios of 3 – 20. Nanowires are wire-like nanomaterials with larger aspect ratios than nanorods and usually nanowires exhibit curly morphology, although there is not a clear standard to distinguish nanorods and nanowires. Some nanowire materials have ribbon-like, belt-like or whisker-like morphology, and they are called nanoribbons, nanobelts, and nanowhiskers with a more precise definition. Nanotubes are a kind of 1D nanofiber material with hollow-structures, although the length and aspect ratio vary.

Many synthesis methods have been developed for generating nanofiber materials with various structures. Among these methods, vapor-liquid-solid (VLS) process, vapor-solid (VS) reaction, laser ablation, solution-liquid-solid (SLS), hydrothermal/solvo-thermal sol-gel and electrospinning routes have been widely employed. Other techniques make use of a straight nanochannel structure, such as SBA-15 and porous anodic aluminum oxide (AAO), as hard template, or deploy liquid crystal and surfactant as soft template. The recent advances in synthesis and development of nanofiber materials and the potential applications of these 1D nanostructured materials are summarized in this book. Potential hazard associated with the research and practical application of nanofiber materials are also addressed. The corresponding materials are organized into three main sections: (1) Preparation methods, (2) applications of nanofibers and (3) risk assessment of nanofiber materials toward application.

Chapter 1

Preparation Methods of Nanofibers

The formation of a 1D nanostructure generally deploys a "bottom-up" stratagem to direct the growth of nanofibers in preferential direction from vapor, solution and solid phase [1,2]. The methods being used for fabrication of various nanofibers in recent years are summarized here.

1.1. Metal-Catalyzed Vapor-Liquid-Solid (VLS) Deposition Method

The mechanism of the VLS process had been proposed by Wagner and Elli in 1964 and the mechanism has been expanded in the followed studied [3,4]. This route has been the prevalent approach for preparation of various nanofiber materials and it is still widely used for fabricating nanofibers of semiconducting materials, where the preparation parameters are modified and the process is incorporated with other techniques for improvement in quality and cost-saving of products. In a typical VLS deposition route, as shown in Figure 1, the synthesis is usually performed at high temperature to create a vapor of target solid materials. Nanofibers grow in the presence of impurity as catalyst to form a liquid alloy nanosize droplet of relatively low freezing temperature. This liquid droplet is a preferred site for deposition of materials from the vapor, which cause the liquid to be supersaturated with target composition. A whisker grows by precipitation of target material from the droplet and the continuous growth resulting in formation of nanorods, nanobelts, nanowires or nanotubes in solid state. The essence of nanofiber formation from this route is a crystallization process from vapor catalyzed by impurities of metal or oxides. By controlling the VLS condition, the obtained nanofibers could exhibit different morphologies.

A large number of nanofiber materials have been synthesized via high temperature VLS routes. This VLS method is specially suitable for the preparation of semiconductor 1D nanostructures, such as Si [5,6,7], ZnO [8], GaN [9] GaAs[10], SnO_2 [11,12], In_2O_3 [13], Bi_2O_3 [14] and others. In most cases, this approach needs the presence of a metal catalyst, such as Fe, Au, Ag, Ga or Sn, to initiate the nucleation of 1D nanostructure at the process temperature usually above 1000 °C [1615,16,17,18]. Silicon nanowires could be grown with Fe catalyst at 1100 °C [19], and the silicon oxide nanowires and nanotubes were grown at 1350 °C on Fe-Ni-Co alloyed catalysts [20]. Hierarchical structures of single-crystal Si nanowires standing on SiO_2 microwires have been synthesized by using Sn catalyst at 1200°C [21]. Much effort has been made to lower the temperature for synthesis of nanofibers via the VLS route. Nguyen *et al.* [22] investigated period

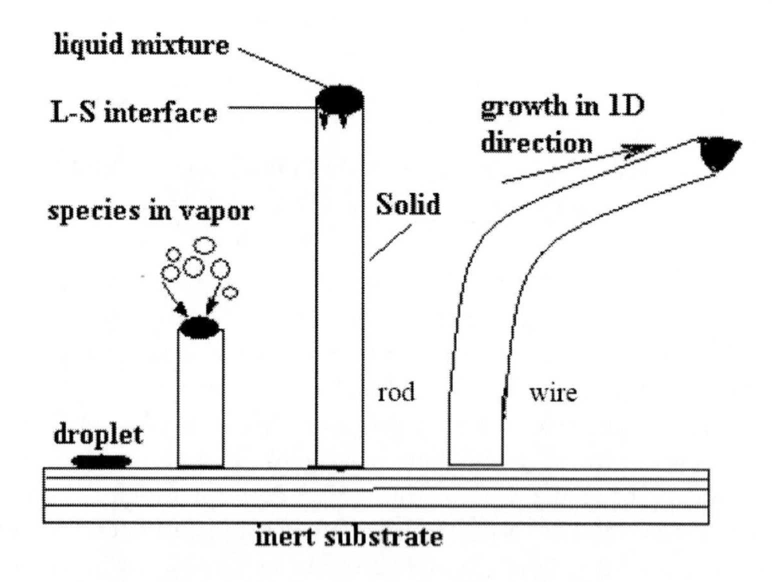

Figure 1. Scheme of nanofiber growth through VLS mechanism.

IV metal elements (titanium, chromium, iron, cobalt, nickel, and copper), period V metals (niobium, molybdenum, palladium, and silver) as well as group-VI and group-III metals as catalysts for tin oxide nanowire growth at temperatures of 840-860 °C. As shown in Figure 2, the morphology of nanofibers was significantly affected by the metal catalyst used. Using titanium catalyst, high-density growth of tin oxide nanoneedles and nanowires were obtained on the substrate surface. The nanoneedles appeared to have a triangular base (~90 nm on a side), which tapered toward the growing front, and an average length of 1.6 μm. The typical nanowire had a diameter of ~ 40 nm, which was uniform throughout the length of the wire (2.2 μm). As a comparison, on catalyst of iron, nanobelts (-210nm × 35 nm × 10 μm) were formed on the substrate. Dense nanowire growth was observed on the copper catalyst. The nanowires had an average diameter of ~85 nm and an average length of more than 10 μm.

Among all metal catalysts, gold has been dominated as the catalyst of choice for inorganic nanowires synthesis through the VLS process due to its chemical inertness and thermal stability. Silicon nanowires were synthesized by gold-catalyzed growth [23-25]. Morral *et al.* [26] recently reported that silicon nanowires were synthesized by using a VLS method using gold nanoparticles as catalyst. The substrates used were thermally oxidized Si substrates with an oxide thickness of 1 μm, which were covered with a 2–3 nm thin gold film by sputtering. The substrates were introduced into a vacuum furnace (2×10^{-6} mbar) and heated to a deposition temperature between 500 and 650 °C with a constant flow of 100 sccm of hydrogen to ensure thermal equilibrium. During the heating process, gold nanoparticles of sizes ranging from 10 to 100 nm were formed from the original gold thin film. For gold-catalyzed silicon nanowires growth, 2.5% of silane was added to the hydrogen flow in the furnace. The total gas pressure was controlled at 10 mbar. Figure 3 displays the SEM images of formed silicon nanowires. A gold catalyst tip was clearly shown on a silicon nanowire, indicating that the growth of silicon nanowires followed the VLS mechanism.

In addition to silicon nanowires, many kinds of semiconducting 1D nanostructures were synthesized by using gold catalyst [27,28]. Well-aligned β-Ga_2O_3 nanowires were synthesized on Au-pre-coated sapphire substrate through VLS route at a low temperature of 550°C [29]. In the presence of oxygen, In_2O_3 nanofibers have been found to be synthesized by a thermal evaporation-oxidation of InP and gold nanoclusters played a crucial role in directing the growth of In_2O_3 nanofibers based on VLS mechanism [10]. With the presence of gold catalyst, large scale CdS nanowires were achieved by thermal vaporization of CdS powders and deposition under controlled conditions [30]. The synthesized CdS nanowires have length up to several tens of microns and diameters about 60-80 nm. This gold-catalyzed growth of CdS nanofiber through VLS mechanism could be generalizable as a means of fabrication of II-VI semiconductor nanowires.

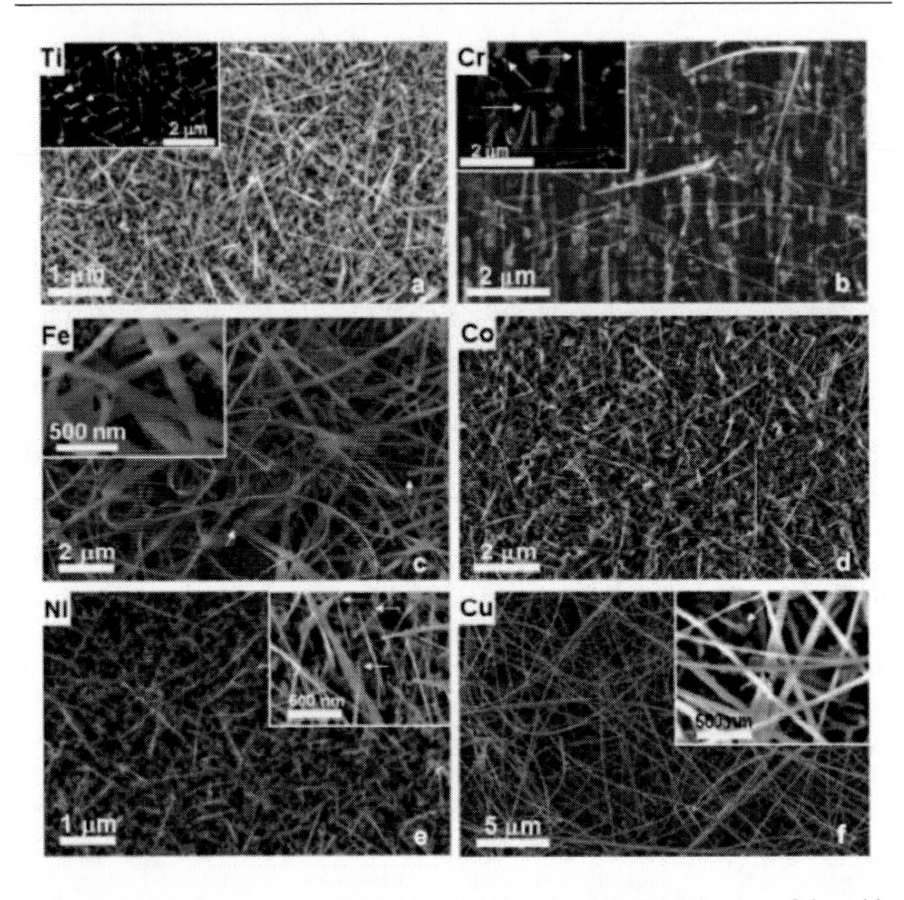

Figure 2. Field-emission scanning electron microscopy (FE-SEM) images of tin oxide nanowires growth on an α-sapphire substrate using group IV metal catalysts. (a) aperspective of high-density tin oxide nanowires using Ti catalyst. The inset shows a higher magnification of the nanowires, showing the presence of both nanoneedles and nanowires at a shorted growth time of 15 min. (b) typical nanowires grown using chromium as catalyst. The higher-magnification inset shows unidirectional island formation and nanowire growth both on the island and on the substrate surface. (c) nanobelt formation using iron as catalyst. The inset shows a higher-magnification image. (d) A perspective view of high-density tin oxide nanowires grown using a cobalt thin film, showing both nanoneedles and nanowires. (e) High-density tin oxide nanowires on a sapphire subatrate, grown using a nickle thin film catalyst. The inset shown a higher magnification of the nanowires and the catalyst terminals of nanowires. (f) High-density and elongated tin oxide nanowires, grown using a copper thin film. The inset shows a higher magnification of the nanowires and catalyst particles at the growing front. [reprinted with permission from Ref 22: Nguyen *et al.*, Adv. Mater., 2005, 17, 1773-1777, Copyright 2005 Wiley-VCH Verlag Gmbh & Co. KGaA, Weinheim].

Figure 3. General morphology of silicon nanowires: a) Typical SEM image of the as-grown silicon nanowires, b) Bright-field TEM and c) energy-filtered TEM images of a typical Silicon nanowires. (c) was obtained by filtering the energy corresponding to the Si plasmon (17 eV). [reprinted with permission from Ref 26 : Morral, et al Adv. Mater., 2007, 19, 1347–1351, © Copyright 2007 Wiley-VCH Verlag Gmbh & Co. KGaA, Weinheim].

Of great interest, Bi_2O_3 nanowires were synthesized by self-catalyzed of Bi and oxidation reaction-coupled VLS route [14]. Bismuth metal was thermal evaporated and deposited on substrate as catalyst. Meanwhile, part of bismuth was also oxidized by pulse supplied oxygen, thus nanowires of β-Bi_2O_3 were effectively formed via the self-catalyzed process. An added advantage is that the nanowire synthesis is free of foreign-element catalyst. When the metal droplets at the tips of nanowires were oxidized, pure oxide nanowires could be obtained. This self-metal catalyzed VLS can be applicable for preparation of a number of metal oxide nanowires with high purity [31].

VLS process for fabrication of nanofibers has been developed with combination of other technology to improve the catalyzed growth of 1D nanostructures. Morales *et al.* [15] introduced a method combining ablation cluster formation and VLS growth process for the synthesis of semiconductor nanowires. Laser ablation was used to prepare nanometer sized catalyst cluster that defined the size of wires produced by VLS growth. As indicated in Figure 4, by ablating the target containing $Si_{0.9}Fe_{0.1}$ at 1200 °C, silicon nanowires with remarkably uniform diameter ~10 nm and length >1 μm were obtained. In addition, the TEM images showed that virtually all of the nanowires terminate at one end in nanoclusters with diameters 1.5 to 2 times that of the connected nanowire. The observation of nanocluster spheres at the ends of the nanowires was indication of a VLS growth process.

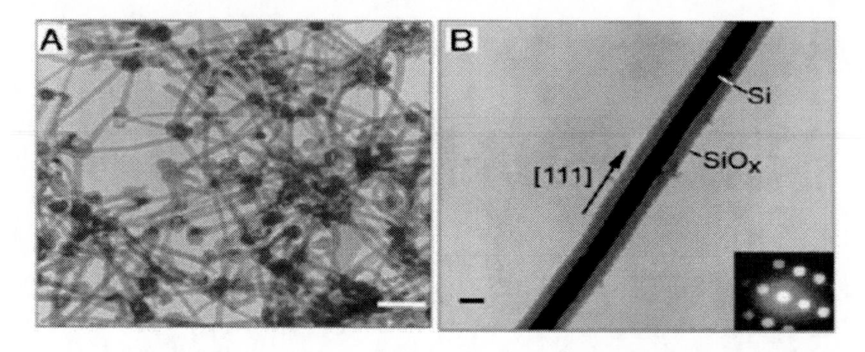

Figure 4. (A) A TEM image of the nanowires produced after ablation of a Si0.9Fe0.1 target; Scale bar, 100 nm. The growth conditions were 1200°C and 500-torr Ar flowing at 50 standard cubic centimeters per minute (SCCM). (B) Diffraction contrast TEM image of a Si nanowire; crystalline material (the Si core) appears darker than amorphous material (SiO_x sheath) in this imaging mode. Scale bar, 10 nm. [reprinted with permission from Ref 15: Morales *et al.*, Science, 1998, 279, 208-211, © 1998 American Association for the Advancement of Science].

A higher resolution TEM image (Figure 4-B) of individual nanowires revealed that the nanowires consist of a very uniform diameter core coating by an amorphous coating.

VLS route combined with Laser-assisted Catalytic Growth (LCG) method was adopted as general synthesis route for fabrication of semiconducting materials of binary group III-V (GaAs, GaP, InAs and InP), II-VI(ZnS, ZnSe, CdS and CdSe) compounds [32]. Target used for synthesis consisted of binary semiconductor components and catalyst element of gold, silver or copper. Under laser ablation, catalyst nanoparticles were formed and directed the VLS growth of crystal nanofibers of target semiconductor. The resulting nanofibers have controllable and uniform diameters. The quality and yield of nanofibers was enhanced by LCG technique. The availability of these high-quality, single-crystal semiconductor nanowires was expected to enable fascinating opportunities in fabrication of nanodevices. Moreover, LCG-VLS approach can be used to synthesize more complex nanowire structures, including single-wire homo- and heterojunctions, and superlattices, and thus may enable the synthesis of nanoscale light-emitting diodes as well as laser devices. In addition, various 1D nanostructure ferrites, such as nanorods, nanowires and nanobelts have been synthesized by using pulsed laser assisted VLS growth and these 1D nanostructured ferrites have great potential in biomedical application [33].

Plasma assisted catalyzed-growth (Plasma-VLS) method was deployed to produce silicon nanowires [34]. By using a 13.6-MHz rf power (10–15W) to create a plasma, and pure SiH_4 as the Si source gas, silicon nanowires grew either by condensing on Au catalyst films, or by self-condensation of the vapor in a lower temperature region of the furnace. The temperature for vapor deposition on gold catalyst was lowered to 300 °C. As shown in Figure 5, with the assistance of plasma, the catalytic growth rate was dramatically enhanced as the length of wires reached tenths of microns and the morphology of silicon nanowires became more uniform.

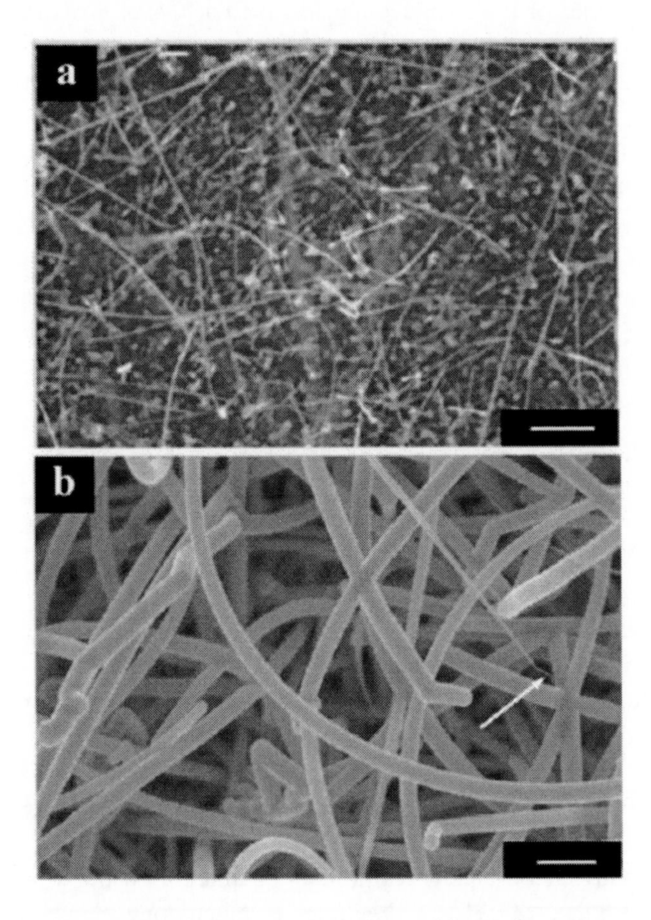

Figure 5. (a) Silicon nanowires grown at 3 Torr without plasma on a uniform Au catalyst layer. Scale bar: 500 nm. (b) Silicon nanowires grown at 3 Torr but with the use of a 10-W plasma. Wires are tenths of microns in length and tapered, as deduced by their larger average diameter and by the presence of sharp tips where the Au catalyst particle can still be found (see arrow). Scale bar: 1 μm. [Reprinted with permission from Ref 34: Colli *et al.*, *Appl. Phys. A,* 2006, 85, 247–253, © 2006 Springer. Part of Springer Science+Business Media].

When nanowires were synthesized at low temperature below the eutectic point of materials-catalyst mixture, the mechanism for VLS growth is disputed. A solid-phase diffusion mechanism was proposed [35]. The target materials from vapor diffused through solid particles of gold seed particles to continue the growth of nanowires of GaAs. This mechanism well explained the plasma-enhanced chemical vapor deposition method to grow Si nanowires at low temperature of 300°C with gold as catalyst. Moreover, gold-catalyzed growth of a series of III-V semiconductor, such as GaP, GaAs, InP and InAs, branched nanowires at low temperature was investigated [36]. The catalytic particles could be in solid phase during nanowires growth. Low-temperature ($T < 400^{o}C$) growth of ZnO nanowires could be achieved in the presence of γ-AuZn particles as catalyst [37].

Although many metal nanoclusters have been proven to catalyze growth of nanofibers via VLS route, this method has an inherent disadvantage. The existence of metal-alloy head at the growing front of nanowires and intrinsic thermodynamic solubility of metal in the nanowires crystal matrix present formidable contamination issues [38]. The metal impurity inevitably influenced the properties of semiconductor nanowires [39,40].

1.2. Oxides Assisted Growth (OAG) of Nanofibers

OAG route is a complementary route to VLS process for the synthesis of nanowire materials. The growth of nanofibers is catalyzed by using oxides instead of metals in VLS process. The OAG of nanofibers involved an oxidation-reduction reaction and the growth mechanism is different from the conventional VLS growth mechanism. Silicon nanowires have been synthesized by laser ablation and evaporation of highly pure silicon powder mixed with SiO_2 at high temperature [41]. SiO_2 was discovered to be the special and effective catalyst which greatly facilitated the growth of silicon nanowires. The silicon nanowires product obtained by using a powder target composed mixture of SiO_2 and silicon is much longer than using metal containing target. As shown in Figure 6, Si nanowires with uniform diameters and smooth surfaces were obtained, while, Si nanoparticles coexisted with the nanowires. Most Si nanowires were extremely long ~10 μm and randomly oriented. It is interesting to note that most Si nanoparticles appear in the form of chain as marked by the arrow.

The mechanism of growth of silicon nanowires was different from that of classic VLS growth [42]. In this OAG procedure, little amount of silicon vapor could be generated at 1200 °C in argon environment. The vapor phase generated by thermal evaporation of solid source ($Si+SiO_2$) mainly consisted of SiO due to reaction between Si and SiO_2. The nucleation of nanoparticles of silicon may involve decomposition of the SiO vapor at a relatively low temperature of 930 °C as: 2SiO (v) → Si (s) + SiO_2 (s). The SiO cluster has been theoretically investigated to be favorable precursor for forming nanostructured silicon, especially for nanofibers of silicon [43]. This decomposition resulted in the precipitation of silicon nanoparticles (nuclei of Si nanowires) surrounded by shell of silicon oxide. The silicon sub-oxide cluster was deposited on the substrate and some of highly reactive silicon atoms were bonded to substrate limiting the motion of the silicon nanoparticle precipitation. Non-bonded active silicon atoms in the same cluster exposed to the vapor and function as nuclei that adsorb addition of reactive silicon oxide cluster. The subsequent growth of the silicon crystal after nucleation may be crystallographically dependent. Oxygen atoms in the silicon sub-oxide clusters might be expelled by silicon atoms during the growth of nanowires and diffused to the edge forming an inert layer of SiO_2 sheath that it prevented the nanowires from growing in diameter. As the nanowires along <112> and <110> have very low surface energy, the silicon nanowires fabricated by the oxide-assisted growth (OAG) are mostly of these two orientations [44]. In addition to wires-like silicon nanofibers, other morphology (such as taper and rod-like) silicon nanofibers were also fabricated through OAG route in nitrogen ambient [45]. The perfect crystalline structure and taper-like geometry of the nanowires rendered the materials excellent field-emission characteristics.

SiO_2 was also found to assist the growth of α-Al_2O_3 fibers at high temperatures [46-48]. In this synthesis, small pieces of Al were placed over a shallow bed of SiO_2 in an inert furnace atmosphere of Ar at temperatures between 1300 and 1500 °C. As shown in Figure 7, the 1D fibers of α-Al_2O_3 with high aspect ratio was obtained after 2 ~ 4 hr of thermal treatment. The mechanism for formation of α-Al_2O_3 fibers with the assistance of SiO_2 might follow this route [47]:

$4Al\ (g) + SiO_2 \rightarrow 2Al_2O\ (g) + Si\ (s,l)$
$2Al(g) + SiO_2 \rightarrow Al_2O\ (g) + SiO\ (g)$
$Al_2O(g) + SiO\ (g) \rightarrow Al_2O_3 + 2\ Si(l)$

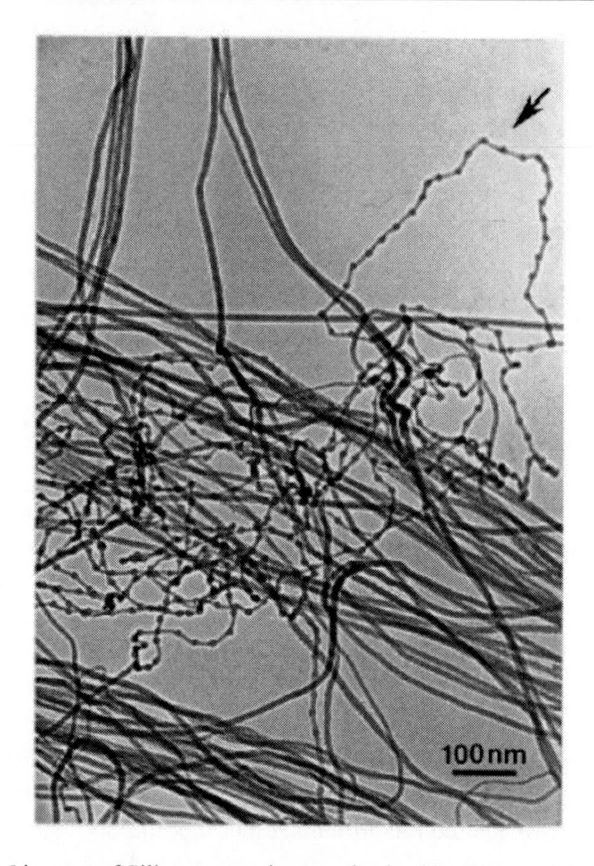

Figure 6. TEM images of Silicon nanowires synthesized by SiO_2-assisted growth. [Reprinted with permission from Ref 41: Wang *et al.*, Phys. Review B, 1998, 58, R16024-16026, © 1998 The American Physical Society].

The gas species, Al_2O and SiO, are continuously being produced by reaction between gaseous Al and SiO_2 powder, while liquid could also be generated owing to eutectic in existence of Si-SiO. Gases dissolved into liquids and react to produce α-Al_2O_3 and the crystal growth in 1D direction to form fibers and ribbons with high aspect ratios. Such features make these fibers suitable as strengthening elements in composite materials.

Conventional OAG is generally carried out in air or low vacuum condition; the products formed in this method could be affected by other impurity in gaseous environment. High quality of silicon nanowires was obtained by OAG process performed at ultrahigh vacuum condition [49]. In addition, the OAG method could be extended for synthesis of a number of nanowires, such as AlN [50], Ge [51], GaAs [52], and ZnO[53].

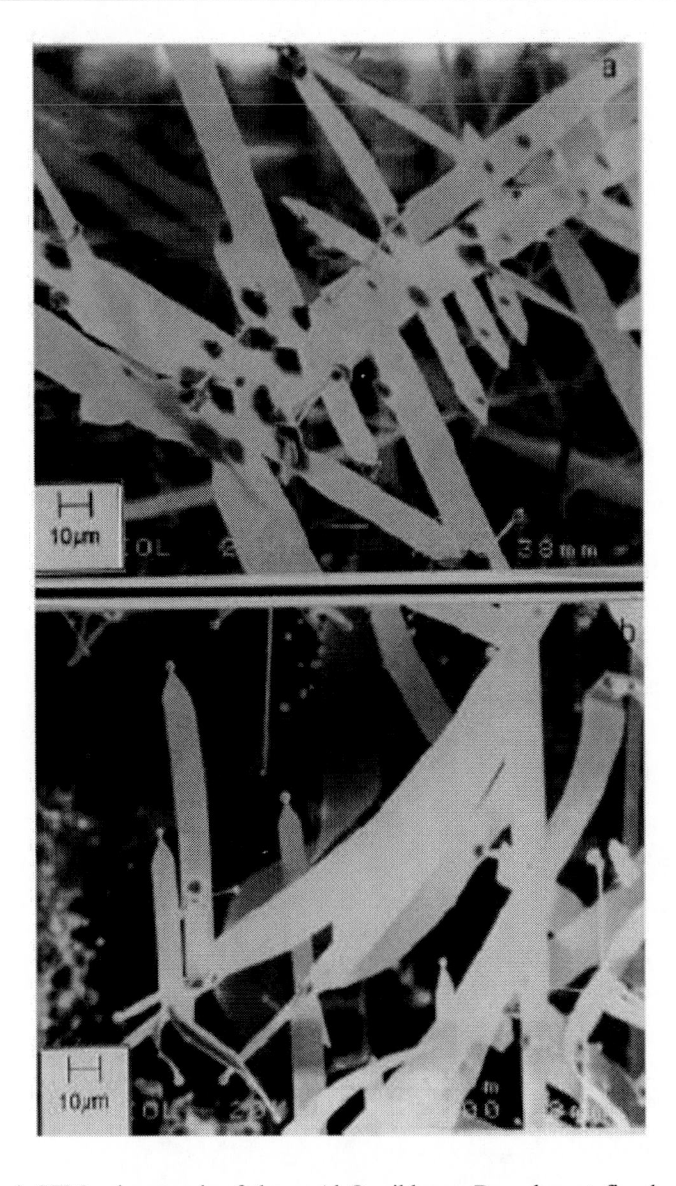

Figure 7. a) SEM micrograph of the α-Al_2O_3 ribbons. Branches at fixed angles and triangular-shaped tips can be observed. Note the α-Al_2O_3 whiskers, with hexagonal section, which we have termed fibers throughout the text, growing on the ribbon's surface. b) Sporadic drops appear at ribbon endings. [Reprinted with permission from Ref 47: Valcµrcel *et al.,* Adv. Mater., 1998, 10, 1370-1373, © 1998 Wiley-VCH Verlag Gmbh & Co. KGaA, Weinheim].

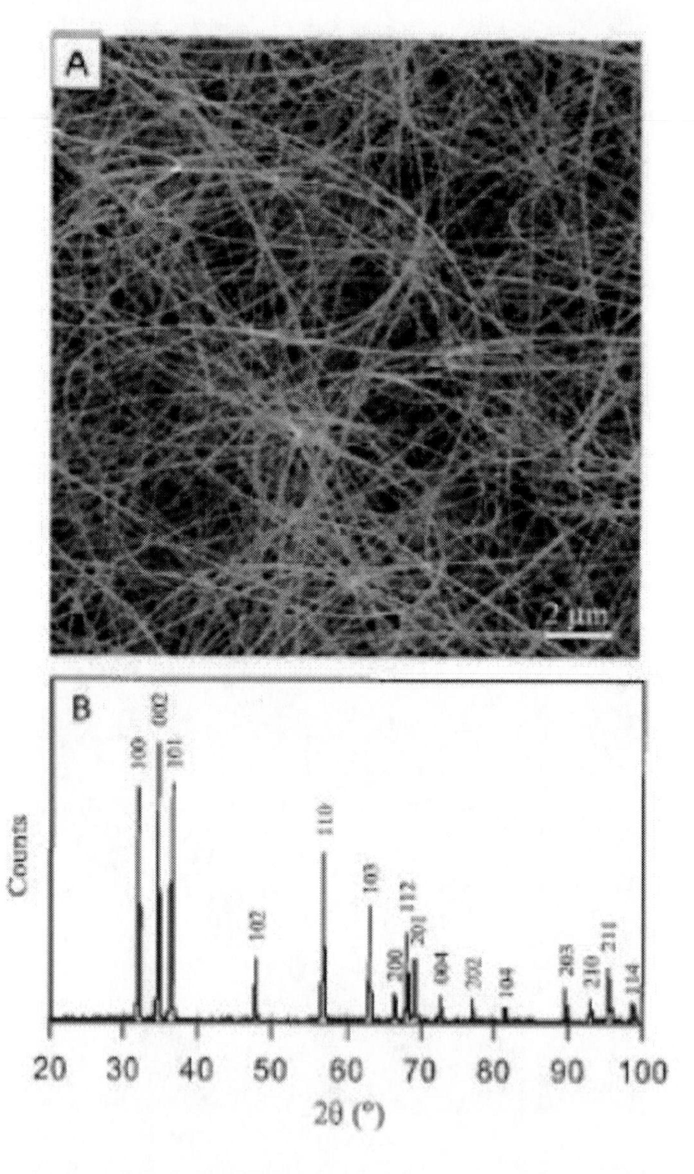

Figure 8. Ultralong nanobelt structure of ZnO (with wurtzite crystal structure). (A) SEM image of the as-synthesized ZnO nanobelts obtained from thermal evaporation of ZnO powders at 1400 °C. (B) XRD pattern recorded from the ZnO nanobelts. [Reprinted with permission from Ref 54: Pan *et al.*, Science, 2001, 291, 1947-1949. © 2001 American Association for the Advancement of Science]].

1.3. Vapor-Solid (VS) Method

Vapor-Solid process is a simpler synthesis method as compared with VLS and OAG routes for fabrication of nanofibers. VS synthesis is based on thermal evaporation of oxide powder under controlled conditions; the vapor was deposited on inert substrate without the presence of any catalyst. The vapor phase could be created by decomposition or reaction, where a chemical vapor deposition (CVD) process is incorporated in the synthesis of nanofibers.

For VS synthesis, normally, the desired oxide powders are placed at the centre of an alumina tube that is inserted in a horizontal tube furnace. The vaporization temperature, pressure and time are controlled subjected to different desired oxide materials. The inert substrate is placed at the downstream of the inert gas flow, where the temperature is lower and the vapor condenses. The 1D nanostructured oxides is formed by crystal growth along the preferential direction.

Ultralong nanobelts of ZnO were synthesized by thermal evaporation of ZnO powder at 1400 °C [54]. After 2 h deposition in Ar flowing with pressure at 300 Torr, high yields of white wool-like products was obtained on alumina plate. As shown in Figure 8, the obtained products were a large quantity of wire-like nanostructure with typical lengths in the range of several tens to several hundreds of micrometers. SEM investigation indicated that this 1D nanostructure was nanobelts having uniform width along the entire length and the typical width of the nanobelts was in range of 50 to 300 nm. Similar morphology of SnO_2 was also synthesized by thermal evaporation at 1350 °C [54]. A host of 1D nanostructured semiconducting oxides, such as SiO_2, In_2O_3, CdO, Ga_2O_3 and PbO_2 were obtained by this VS synthesis route [20, 54], In addition to nanowires of oxides, the direct thermal evaporation-deposition method was used for synthesis of semiconducting nanotubes of Bi_2S_3 [55].

VS synthesis route is normally performed at high temperature to create vapor of oxides. To lower evaporation temperature, a molten-salt-assisted CVD method was used [56]. By mixing Zn powder, NaCl and nonylphenol polyethylene glycol ether, ultralong ZnO nanobelts could be obtained by thermal evaporation and deposition at 800 °C in open-atmosphere environment. The controlled vaporization of Zn from molten NaCl and oxidized to ZnO during deposition allowed ZnO nanobelts to grow on the substrate.

By using $SiCl_4/H_2$ as silicon source in a CVD-VS process, well-aligned Si/SiO_2 composite nanowires were deposited on various substrates at low temperature of 120 °C [57]. Vapor phase silicon source was produced by hot-filament tungsten wire at high temperature. Oxygen source came from the trace amount of moisture adsorbed on wall of the reactor. The oxygen-containing silicon precursor could be formed from $SiCl_4/H_2/H_2O$ in the high temperature region near the hot filament. Due to the high strength of Si-O bond, the precursor would be bonded once they arrive on the nanotips by forming Si-O- bonds. The growth of 1D nanostructure was well aligned as the migration of precursor along the surface is not significant at the low deposition temperature on substrate. The resulting Si/SiO_x ($x = 1 \sim 2$) nanowires contained a silicon core embedded in amorphous SiO_x shell.

A series of 1D nanostructured materials were synthesized by VS process by chemical reaction and deposition on substrate. AlN nanorods were prepared by reaction of $AlCl_3$ and $(NH_4)_2CO_3$ in gas phase at 600°C to form AlN followed by solid deposition on substrate [58]. ε-FeSi nanowires were synthesized by thermal CVD of $FeCl_3$ over silicon wafer in inert environment [59]. The Cu nanowires of 70–250 nm diameter were grown on Si at substrate temperatures of 200–300 °C under 0.1–1.0 Torr using argon as a carrier gas and $Cu(ETAC)[P(OEt)_3]_2$ was used as a precursor [60]. Moreover, ultralong zigzag-shaped Al_2O_3 nanobelts composed of periodically twinned crystal morphology along the nanobelts axial direction through the whole length of belt were fabricated by chemical reaction-coupled thermal evaporation method [61]. The precursor vapor Al_2O was produced by the reaction between Al and Al_2O_3 at a relatively high temperature [62]. Then, when Al_2O was transported to the downstream lower temperature zone in the ceramic tube, the Al_2O decomposed ($3Al_2O \rightarrow Al_2O_3 + 4Al$) or oxidized ($Al_2O + O_2 \rightarrow Al_2O_3$) to Al_2O_3 and deposited on the substrate. The formed Al_2O_3 nucleated and grew along 1D preferential direction to form nanobelts, as indicated in Figure 9. The nanobelts of Al_2O_3 exhibited outstanding dielectric properties as compared to Al_2O_3 micropowder.

Various nanofiber materials were synthesized by VS route, which is either by direct thermal evaporation-deposition or reaction-coupled vapor-deposition (CVD) method as well as plasma-enhanced CVD [63,64]. This process usually need to be performed under high temperature condition which could be free of catalyst or foreign-element, thus target nanofibers with high purity could be achieved. On the other hand, physical vapor deposition (PVD) method could be used for fabrication of organic nanowires at low temperatures [65].

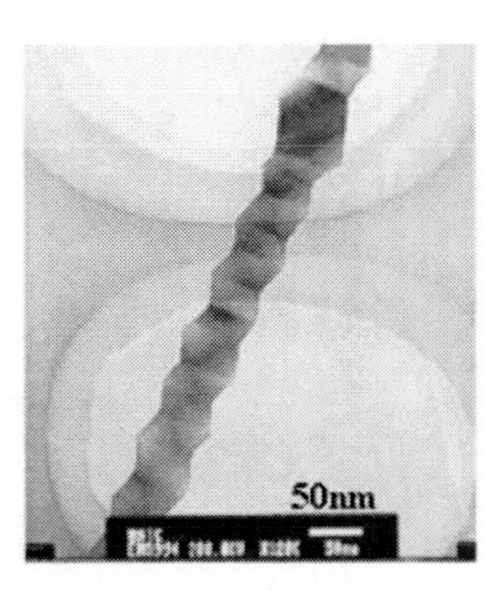

Figure 9. SEM image (left) and TEM image (right) of the zigzag-shaped Al_2O_3 nanobelts. [Reprinted with permission from Ref 61: Fang *et al.*, Adv. Mater., 2005, 17, 1661-1665. Copyright 2005 Wiley-VCH Verlag Gmbh & Co. KGaA, Weinheim].

1.4. Hard Template Directed Synthesis Route

Templated synthesis route is based on available nanofibers, nanotubes or nanostructured materials with 1D arranged nanostructures to synthesis another kind of 1D nanostructured materials. Figure 10 illustrates the synthesis scheme of different nanofiber materials via deigned templated routes. Nanorods and nanowires can be fabricated by filling the target materials into uniform nanopores through solution diffusion or vapor deposition. After removing the hard template by reaction with reagent, which is inactive for filler within nanopores, nanowires or nanorods were obtained depending on the shape and length of nanoporous templates. By deposition of target materials onto the inner surface and the nanopores are not fully filled, nanotubes are obtained after removing the nanoporous template. Nanotubes can also be fabricated by coating a layer on to the outer surface of solid nanofibers and followed by removing the fiber core. The nanowires can be fabricated by assembly of solid nanoparticles or clusters along available nanofibers or linear microorganism under mild conditions. Silica based mesoporous materials with ordered arranged nano size pore channels are a good candidate for synthesis of nanorods, nanotubes and nanowires. Porous anodic aluminum oxide (AAO) membrane has been extensively used as template for fabrication of nanotubes. Several kinds of nanostructured materials used as hard template for fabrication nanofibers with different morphologies followed the different routes will be described below.

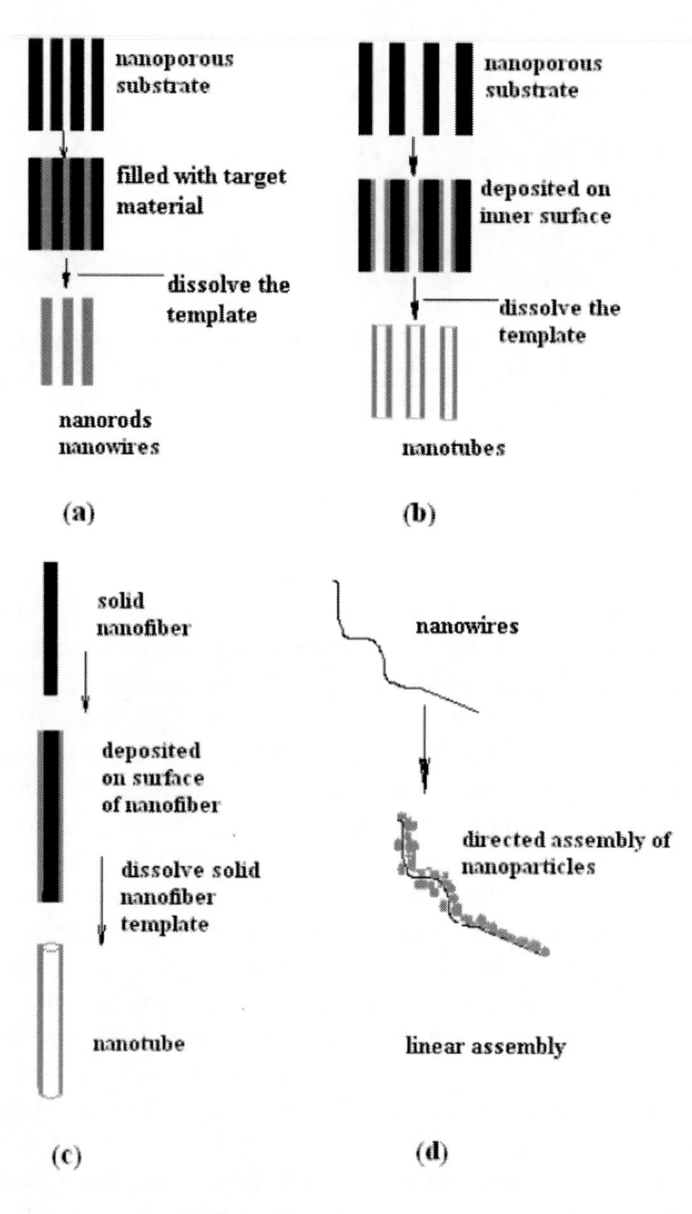

Figure 10. Schematic diagram of different templated-methods for synthesis of nanofiber materials.

1.4.1. Mesoporous Templates

As illustrated in Figure 10(a, b), nanoporous materials are candidates for fabrication of nanorods, nanowires and nanotubes. Among porous materials, ordered mesoporous silica has been discovered and intensively explored for application in recent decade [66]. Due to the uniform pore size and 1D arranged pore channels, mesoporous silica is used as template for fabrication of nanorods or nanowires [67-70]. The diameter of resulting nanorods or nanowires is 6-7.5 nm, which depends on the pore size of mesoporous silica. After removing the template, the ordered arrays of nanofibers will be obtained. A simple impregnation reaction method was used to synthesize nanowires of metal sulfide (CdS, ZnS, and In_2S_3) by filling the nanopores of SBA-15 with two separate precursors (metal and sulfur precursors) [71-73]. The hard template of SBA-15 was easily to be removed as the amorphous silica pore wall can be dissolved in NaOH or HF solution. Metal (Pt, Au, Ag) nanowires were also obtained by solution impregnation method followed by reduction with hydrogen [74]. The deposition method was investigated and found to affect the quality of formed nanofibers. Pulse electrodeposition resulted in ultra high density CdS nanowires arrays [75]. GaN semiconductor nanorods were fabricated inside of nanoscale channels of mesoporous silica SBA-15 [76].

When the target component is coated on the inner pore wall of mesoporous silica and the pores were not fully filled, nanotubes are obtained after removing the template. Boron nitride nanotubes were obtained by CVD of BCl_3 and NH_3 and reaction into pore channel of SBA-15 and followed by dissolving the silica template wall with HF solution [77]. Carbon nanotube arrays were formed by catalytic decomposition of acetylene at 700 °C inside the pore channels of SBA-16 [78]. The carbon nanotubes are very uniform in diameter and length and are aligned vertically with respect to the SBA-16 film. Uniform carbon nanowires with diameter below 1 nm was developed using a confined self-assembly in MCM-41 as hard template [79]. In addition to silica based mesoporous materials, nanoporous polycarbonate membrane could also be used as template for fabrication of oxide nanotubes and semiconductor nanowires [80]. ZrO_2 and TiO_2 nanotubes were synthesized by using polycarbonate nanoporous membrane as template [81]· Precursor was coated onto inner surface and porous template was removed by etching with chloroform. Moreover, a series of single-crystalline alkaline-earth metal fluoride (such as CaF_2, SrF_2 and BaF_2) nanowires were synthesized using polycarbonate membrane at ambient condition [82]. When polycarbonate was immersed in mixed solution of NH_4F and $CaCl_2$ (or SrC12, $BaCl_2$), the confined crystal growth of metal-fluoride inside track-etched channels was performed at room temperature. After removing polycarbonate template, uniform nanofibers of alkaline-earth metal fluoride were obtained.

1.4.2. AAO Templates

AAO membrane is a kind of commercialized products (Whatman Inc Co.) with various pore sizes in the range of tens to several hundred nanometers, which could be prepared by anodization process of aluminum containing alloy or aluminum membrane in phosphoric acid solution and followed by a thermal anneal [83,84]. AAO membrane has been widely used as templates for fabrication of nanotubes, especially for formation of metal nanotubes. A facile method was conducted to synthesize silver nanotubes by filling the nanopores with $Ag(NO_3)$ solution through capillary interaction [85]· After drying, solid $Ag(NO_3)$ adhered to the pore wall of template. Upon thermal treatment at 500 °C, $Ag(NO_3)$ on the inner pore wall decomposed to metallic silver and gaseous products to form Ag nanoparticle nanotubes. Finally, as shown in Figure 11, silver nanotubes were freed from the template by dissolving the alumina membrane with 0.1M KOH aqueous solution and then washed with distilled water [86]. This chemical-free facile route is applicable for fabrication a large number of stable metal nanotubes, such as Pt [87], Pd [88,89], Ru [90] and Au [91,92] tubes. These metal nanotubes exhibit good electric conductivity as they are not coated with organic layer.

Metal nanoclusters or nanoparticles were immobilized onto the inner wall of AAO template by various methods. For molecular anchors binding the metal nanoparticles, the surface of AAO channels was modified with organosilane molecules for self-assembly of Au inside the channels [93]. Rubinstein and co-workers demonstrated the preparation of Au nanoparticle nanotubes by introducing a colloid solution of Au nanoparticles into the pores of an organosilane-modified AAO, followed by spontaneous coalescence of the surface-bound Au nanoparticles [94]. Alternatively, Pd nanoparticles could be effectively deposited on the pore walls of AAO by using a sequential electroless deposition technique, in which a palladium complex ($[Pd(NH_3)_4]^{2+}$) was thermally reduced to metallic Pd [95]. Au nanotubes were also prepared by radiofrequency (RF)-sputtering of Au particles through a template-directed synthesis in porous alumina substrates.[96] Electrodeposition of Au and Ni source into AAO pore channels with controlled cycles, multisegmented metal nanotubes was obtained (as indicated in Figure 12) after removing the AAO template [97,98]. This nanofabrication method can be readily extended to a wide range of metallic or semiconducting materials. Composite nanoparticle assembled nanotubes of Au-Pd were obtained by using mixed solution and deposited into AAO template [99]. Formation of intimate metal-metal interfaces by room-temperature coalescence of different kinds of metal particles is an approach to the synthesis of nanostructured materials under mild conditions using metal nanoparticles as building blocks.

Figure 11. SEM images of silver nanotubes with different orientations and at different magnifications. [Reprinted with permission from Ref 85: Qu *et al.*, Adv. Mater., 2004, 16, 1200-1203, Copyright 2004 Wiley-VCH Verlag Gmbh & Co. KGaA, Weinheim].

Figure 12. SEM images of multisegmented metal nanotubes with a stacking configuration of Au-Ni-Au-Ni-Au along the nanotube axis. a) Cross-sectional SEM image of as-prepared metal nanotube–AAO composite, which shows metal nanotubes embedded in an alumina matrix. The signals from Au and Ni are shown in yellow and purple, respectively. b) and c) SEM images of multisegmented metal nanotubes after removal of alumina matrix with NaOH (1.0m); part c) clearly shows the stacking configuration of multisegmented metal nanotubes in which the segments with bright and dark image contrasts correspond to Au and Ni, respectively. [Reprinted with permission from Ref 97: Lee *et al.*, Angew. Chem. Int. Ed., 2005, 44, 6050 –6054].

In addition to metal nanotubes, oxides nanotubes could be fabricated via the AAO templated route. Similar to procedure for fabrication of metal nanotubes, a convenient method led to the formation of RuO_2 nanotubes derived from $Ru_3(CO)_{12}$ clusters in AAO template [100]. $Ru_3(CO)_{12}$ in hexane solution was introduced to AAO template by wet impregnation. After drying and thermal treatment at 600 °C under nitrogen flow, RuO_2 nanotubes with length up to 3μm were obtained. Moreover, AAO templates have been used for synthesis of titania [101] and carbon [102,103] nanotubes. Interestingly, alumina nanotubes/nanowires can also be obtained by controlled etching of AAO template directly [104,105].

In addition to nanotubes, nanowires have been synthesized by using AAO templates [106]. By electrodeposition of Au into pore channels of AAO till fully filled and followed by template dissolution, nanowires of Au were obtained [107]. The mechanical properties of the resulting gold nanowires were investigated. It was found that Young's modulus was essentially independent of the diameter of nanowires, whereas the yield strength was largest for the wire with smallest diameter. The strength of nanowires is up to 100 times that of bulk gold materials. A host of nanowires, such as ZnO [108], ZrO_2 [109]Sc@C_{82} [110] $La(OH)_3$ [111], $Ni(OH)_2$ [112], AlOOH [113] and polymer [114] nanofibers, have been fabricated by using AAO template. Moreover, polymer membrane with nano-sized pore structures played similar role as AAO template to direct the growth of oxide nanowires [115].

1.4.3. Nanotube Templated Route

Carbon nanotubes are among the most active research topic in recent decade as reviewed [116,117,118,119,120]. Carbon nanotubes are widely used as hard template for fabrication of other materials with nanostructures of nanotubes and nanowires since the discovery in 1991 [121]. The use of carbon nanotubes as confined template for synthesizing other nanofiber was introduced here for synthesis of other nanotubes via a "tube-to-tube" templated route. Semiconductor nanofibers (GaN) can be prepared through a carbon nanotubes confined reaction [122]. Ga_2O vapor was reacted with NH_3 gas in the presence of carbon nanotubes to form wurtzite gallium nitride nanorods, as displayed in Figure 13. The nanorods have a diameter of 4 to 50 nanometers and a length of up to 25 micrometers. It was proposed that the carbon nanotube acted as a template to confine the reaction, which resulted in the gallium nitride nanofibers having a diameter similar to that of the original nanotubes. The results suggested that it might be possible to synthesize other nitride nanorods through similar carbon nanotube–confined reactions. By reaction of carbon nanotubes with volatile oxide/halide species, 1D nanostructured carbide (such as TiC, NbC, Fe_3C, SiC, and BC) in high yield with typical diameter of 2-30 nm were obtained [123]. Similarly, GaP nanorods were synthesized by reaction of Ga_2O in a phosphorus vapor atmosphere [124]. Metal nanowires can be synthesized by filling or external deposition on carbon nanotubes [125].

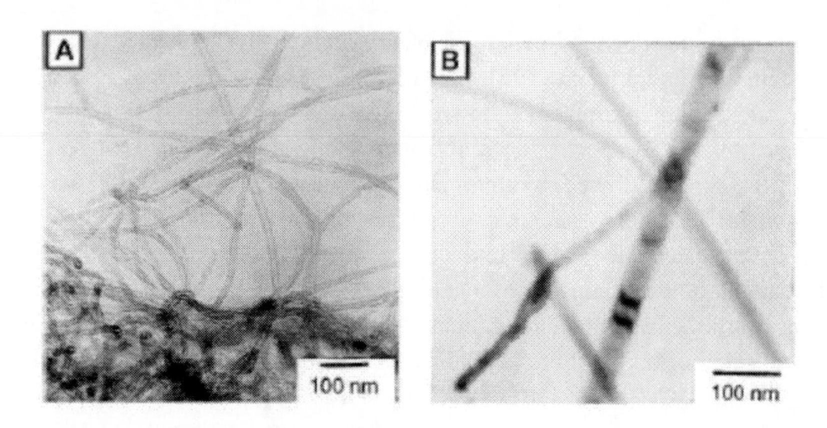

Figure 13. (A) TEM image of the carbon nanotubes used as starting material. (B) TEM image of the GaN nanorods that were produced. [Reprinted with permission from Ref: 122: Han *et al.*, Science, 1997, 277, 1287-1289. © 1997 American Association for the Advancement of Science].

Metal oxide nanotubes were fabricated using carbon nanotubes via a "tube-to-tube" templating route. For synthesis of α-Fe_2O_3 nanotubes [126], carbon nanotubes were first fully coated with iron oxide nanoparticle by thermal decomposition of $Fe(NO_3)_3$ in super critical CO_2-ethanol solution at 150 °C. The resulting composites of Fe_2O_3 and carbon nanotubes were then heated in an oxygen-rich atmosphere to remove the carbon nanotube template, and α-Fe_2O_3 nanotubes were obtained. The as-prepared α-Fe_2O_3 nanotubes exhibited outstanding sensitivity and excellent selectivity to hydrogen sulfide (H_2S) based on the catalytic chemiluminescence, making it an attractive material as H_2S sensor.

As illustrated in Figure 10(d), nanoparticles can be directed to assemble along carbon nanotubes. The surface of nanotubes could be modification by functional materials and create affinity to target nanoparticles, thus the nanoparticles can be assembled along the nanotubes to form 1D arrays. By functionalizing nanotubes with negatively charged polymer wrapping technique [127], the linear assembly of silica-coated Au nanoparticles on carbon nanotubes could be reached by simple procedure [128]. On polystyrene sulfonate wrapped carbon nanotubes, a linear colloid nanocomposite with high aspect ratios, well defined space between neighboring silica coated Au nanoparticles were obtained by layer-by-layer deposition, as shown in Figure 14. The composite nanowires are optically labeled and have potential application as components of nanoelectronic circuits and waveguides.

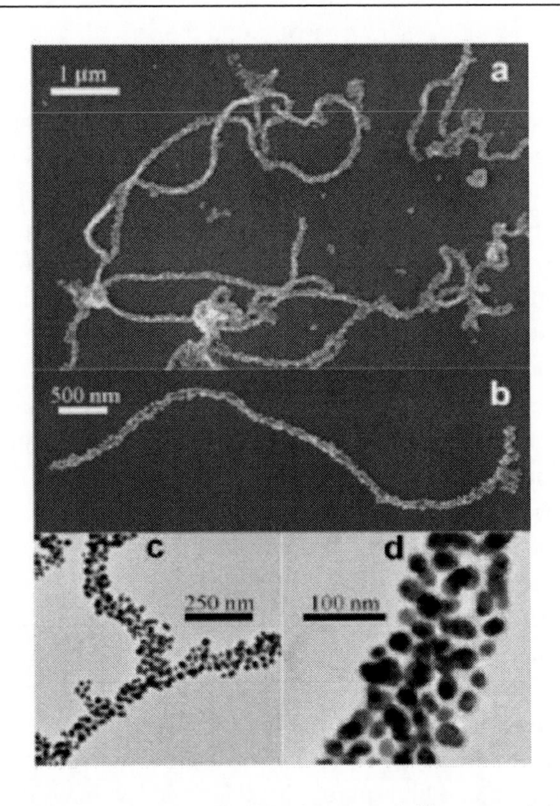

Figure 14. (a,b) SEM images and (c,d) TEM images of silica-coated Au nanoparticles assembled on thin carbon nanotubes. [Reprinted with permission from Ref 128: Correa-Duarte et al, Adv. Mater., 2004, 16, 2179-2183. Copyright 2005 Wiley-VCH Verlag Gmbh & Co. KGaA, Weinheim].

A variety of nanowires were self-assembled along existing nanowires materials as template via the "wires-to-wires" or "tubes-to-wires" routes [129]. Co_3O_4 nanowires and Au-Co_3O_4 hybrid nanowires were fabricated based on virus template and exhibited application as flexible lithium ion battery electrodes [130]. Semiconductor nanoparticles chains and wires could be fabricated by DNA templated route. CdS and SiN nanowires were obtained under a mild condition based on DNA templated fabrication [131,132,133]. Organo-silica hybrid nanowires were reported to be synthesized by the use of core–shell cylindrical polymer brushes as soft templates [134]. Continuous and segmented polymer/metal oxide nanowires were fabricated by using cylindrical micelles and block co-micelles as templates [135]. The highly uniform nanowires with controlled morphology have exhibited the application as nanodevices components.

1.4.4. Nanofiber Templated Synthesis of Nanotubes

Nanotube materials can be fabricated using solid nanofibers (nanorods or nanowires) as template. As illustrated in Figure 10(c), a thin layer of target material is uniformly coated on the nanofibers through various approaches. After removing the fiber core, the preserved shell led to formation of hollow nanostructured fibers, i.e nanotubes. One of the major difficulties for this process is to control the uniformity and dimension of the final product. This process often involves a physical vapor deposition (PVD) [136], chemical vapor deposition (CVD) [136] or sol-gel [137] technique for coating specified target materials. Each of these deposition methods suffered significant limitation to reach the objectives. For the instance, PVD process does not permit conformal coating because it is a direct line-of-sight deposition. Non-uniform deposition usually occurs in the CVD process because of rapid consumption of the precursor. The substrate for the sol-gel coating needs uniform wetting that is usually difficult to achieve.

Among various methods, atomic-layer deposition (ALD) is the most suitable approach route to deposit target materials on the surface to form a conformally uniform layer [138,139]. Al_2O_3 nanotubes were fabricated by using ZnO nanowires as template [140]. The ZnO nanowires were covered with Al_2O_3 by atomic-layer-deposition at a growth temperature of 300 °C. Trimethylaluminum and high-purity distilled water were used as precursor for Al_2O_3 formation and deposition on surface of ZnO nanowires. The thickness of the Al_2O_3 shells was precisely controlled by the time of the cycles of deposition. The core material in the Al_2O_3-coverred ZnO wires were subsequently etched-up in H_3PO_4 solution to yield Al_2O_3 nanotubes, as shown in Figure 15 [141]. Nanotubes can also be synthesized by using organic nanowires as template through ALD route. Al_2O_3 nanotubes were fabricated over template of Tris-(8-hydroxyquinoline) gallium (GaQ_3) organic nanowires [142,143]. After Al_2O_3 was deposited on the organic nanofibers by ALD, the core fiber was removed by thermal treatment at 900 °C. The resulting Al_2O_3 nanotubes exhibited thermal stability and a uniform morphology.

Shape-controlled nanotubes of oxides, such as ZrO_2, Al_2O_3, and SiO_2, can be synthesized by using carbon solid nanofibers of various morphologies [144]. The morphology of oxide nanotubes was controlled by the shape of the carbon nanofibers used. Mono and binary transition metal oxide nanotubes could be synthesized by the immersion of carbon nanofiber templates into metal nitrate solutions and removal of the templates by heat treatment in air [145]. The precursor was deposited onto surface of carbon nanofibers by quick impregnation. For example, precursor diluted with organic solvent ($(Zr(O^nPr)_4$ (150 mM) in C_2H_5OH, $Al(O_{sec}Bu)_3$ (150 mM) in CCl_4, or $SiCl_4$ (500 mM) in CCl_4) was dropped into the carbon nanofiber templates and the excess precursor solution was removed by filtration immediately. Obtained samples were dried in air at room temperature to remove solvent. The precursor remained and was adsorbed on the surface of the templates and was immediately hydrolyzed by the water vapor in air. Consequently, the carbon nanofiber templates were covered with a thin oxide and hydroxide. This coating process was repeated for 10-40 times. The carbon nanofiber templates were removed by calcination in air at 1023 K for 4 h.

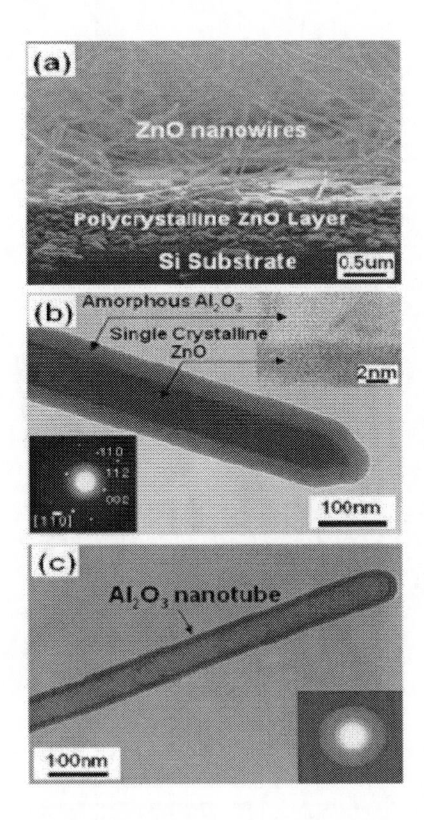

Figure 15. SEM image of (a) ZnO nanowires grown from a thick polycrystalline ZnO layer on a silicon substrate, (b) TEM image of ZnO/Al_2O_3 core/shell nanofiber and (c) an Al_2O_3 nanotube. The top-right inset in (b) shows a HRTEM of the ZnO/Al_2O_3 core/shell nanofiber, bottom-left inset in (b) shows the SAED pattern of the ZnO/Al_2O_3 core/shell nanofiber and the inset in (c) shows the SAED pattern of the amorphous Al_2O_3 shell. [Reprinted with permission from Ref 141: Hwang *et al.*, Adv. Mater., 2004, 16, 422-425, © 2004 Wiley-VCH Verlag Gmbh & Co. KGaA, Weinheim].

As shown in Figure 16, the obtained thermally stable oxide nanotubes exhibited replicated morphology of the carbon nanofiber template. Using similar method, metal nanowires could also be fabricated by carbon nanotubes templated route. After removing the core carbon nanofibers by oxidation in air, the nanotubes were treated in hydrogen to reduce oxide to metal and tubular nanostructure was preserved [146].

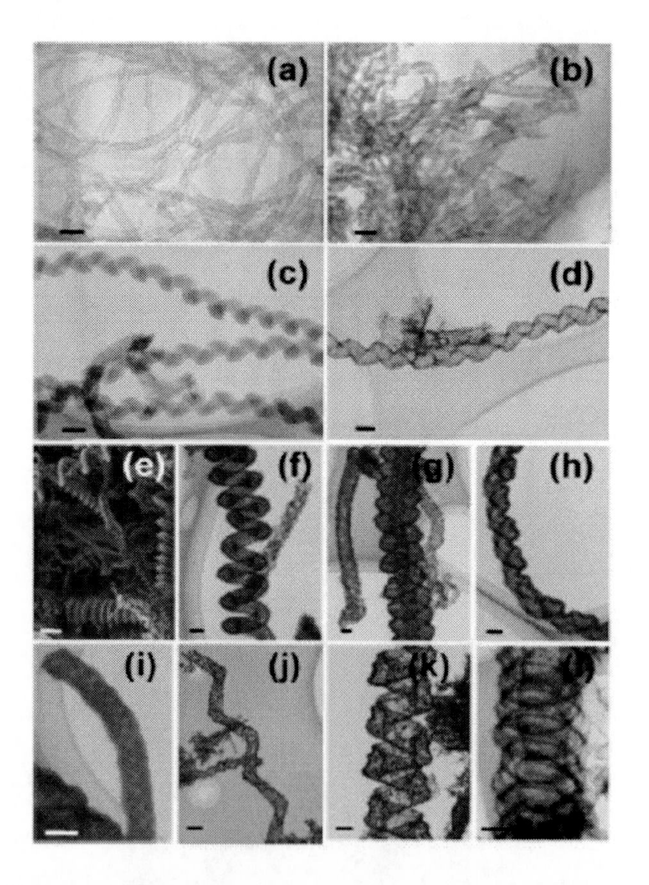

Figure 16. TEM and SEM images of (a) carbon nanotubes, (b) ZrO_2 nanotubes synthesized using template of (a) by 40 coating runs, (c) carbon nanotubes-1, (d) ZrO_2 nanotubes synthesized using template of (c) by 40 coating runs, (e) carbon nanotubes-2 (f-i) SiO_2 nanotubes synthesized using template-(e) by 20 coating runs. (j and k) ZrO_2 nanotubes synthesized using template-(e) by 10 coating runs, and (l) Al_2O_3 nanotubes synthesized using template –(e) by 20 coating runs. The scale bars shown in panels a and b; e; and c, d, and f-l represent 30 nm, 1 μm, and 200 nm, respectively [reprinted with permission from Ref 144: Ogihara *et al.*, Chem. Mater., 2006, 18, 4981-4983. © 2006 American Chemical Society].

The "fiber-to-tube" template route is applicable to be expanded to a variety of semiconductor and oxide nanotubes by sacrificial nanofiber templates [147,148,149,150] or direct reaction with nanofibers [151,152,153].

Generally, the solid template route for fabrication of 1D nanostructure is a nano-manipulated process to produce nanowires or nanotubes with designed shape, diameter or thickness. Template with nanostructure of pore channels or nanofibers simply serves as platform to confine different materials within channels or support on surface to shape the target materials into nanostructure with morphology controlled by nanostructured template used. The precursor of target materials are usually introduced to the template by wet impregnation, PVD, CVD and ALD methods, and it is necessary to selectively remove template by post-synthesis treatment process which is safe to the target 1D nanostructures. Template is usually digested by chemical etching or thermal oxidation, resulting in formation of desired 1D nanostructure. To completely remove template in this templated route is a challenge and the impurity may exist in the final product due to the residual of template and the chemicals used for template etching. Moreover, for synthesis of nanofibers confined in the 1D nano-channels template, it is difficult to fully fill the nano-channels with target solid materials, thus making it challenging to fabricate nanofibers with large aspect ratio and maintain the uniform continuity along the final fibers. When template used as reactant for in-situ production of 1D nanostructure, template material is usually consumed as the reaction proceeds, this route may directly produce the 1D nanostructures as a pure product. It should be noted that the nanowires and nanotubes synthesized by hard-template route are usually polycrystalline or even in amorphous state. In addition, the harsh chemical conditions used for digestion of hard template may lead to environmental impaction, and production scale using hard template route is relatively limited. Thus, more efficient processes for synthesis of nanofibers in terms of cost and potential for large-scale production are desired, as the large-scale manufacture of nanomaterials at an affordable cost stands the ultimate challenge for application of nanotechnology.

1.5. Wet-Chemical Synthesis Methods

Many of the earliest syntheses of 1D nanostructure solid were achieved through liquid solution based routes with advantages of simplicity and suitability for large scale production. Wet-chemical synthesis routes include hydrothermal, solvothermal and sol-gel processes. Nanofibers can be fabricated in liquid phase in the presence or absence of surfactants. Many kinds of nanofibers of hydroxides were grown by controlled crystallization in liquid phase under hydrothermal conditions, and thermal decomposition lead to formation of oxide nanofibers while some kinds of oxide nanofibers were directly formed in liquid phase. Usually, metal nanowires are fabricated in the presence of reductants in solution.

1.5.1. Hydrothermal Synthesis Route

Hydrothermal synthesis is a cost-effective technique for crystallizing substance from aqueous solution under pressure created by heating the closed vessel at a certain temperature. A hydrothermal crystal growth usually results in well crystalline materials, or single crystal in many cases. A large number of compounds with 1D nanostructures, such as simple and complex hydroxides [154, 155, 156], oxides [157, 158], metals [159], sulfates [160, 161, 162, 163], tungstates [164, 165, 166, 167], molybdates [168], binary semiconductors [169, 170, 171, 172] and inorganic-organic hybrid nanofibers [173] have been synthesized under hydrothermal conditions.

Many kinds of 1D nanostructured hydroxide compound, such as AlOOH [174, 175, 176], FeOOH [177, 178], $Ni(OH)_2$ [179, 180, 181], $Ce(OH)_3$ [182, 183], $La(OH)_3$ [184, 185] and MnOOH [186, 187] have been synthesized through a facile hydrothermal route. For example, recently, $La(OH)_3$ nanobelts [188] were reported to be synthesized by simply adding $La(CH_3COO)_3$ to a mixture of hydroxides (NaOH/KOH= 51.5:48.5) aqueous solution in a covered Teflon vessel and heating the mixture in a furnace at 200 °C for 48 h. As shown in Figure 17, nanobelts of $La(OH)_3$ with large aspect ratio was obtained. XRD pattern indicated that the uniform nanobelts were well crystallized and EDS spectrum showed the materials had high purity of $La(OH)_3$. It was also noted that the morphology of nanobelts was well preserved when $La(OH)_3$ was converted to La_2O_3 by a thermal decomposition.

Various nanofibers of hydroxide materials exhibited similar character as $La(OH)_3$ in preserving 1D nanostructure after being converted to oxides [174,177,182,183,187,189,190]. Thus, this method is widely adopted for synthesis of metal oxide nanofibers.

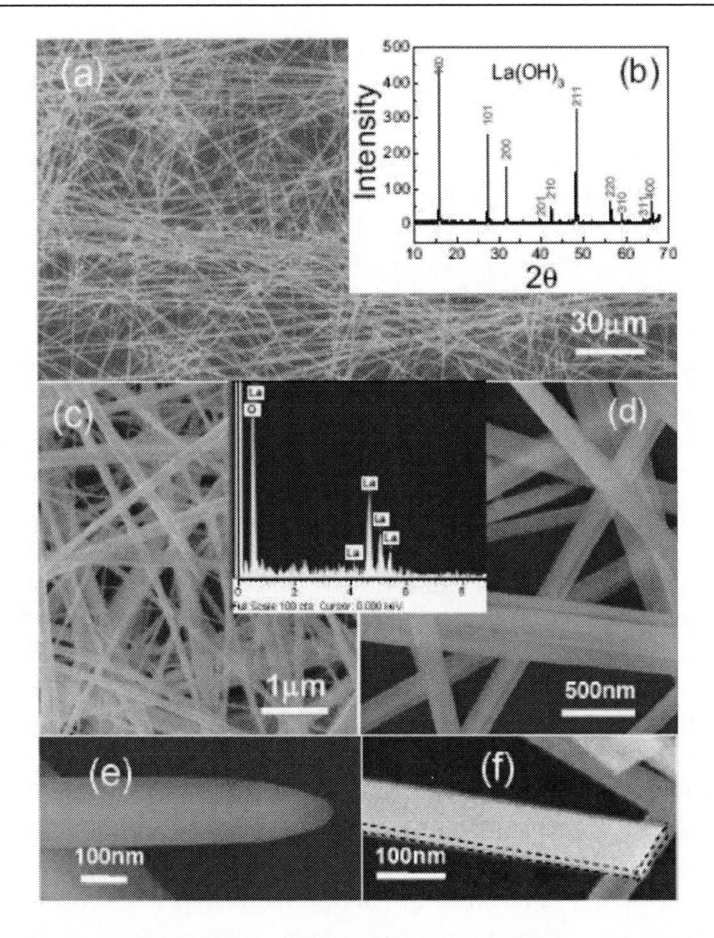

Figure 17. a) Low-magnification SEM image of the $La(OH)_3$ products, indicating lengths of up to several millimeters. b) A typical XRD pattern of the as-synthesized La(OH)3 product. c,d) High-magnification SEM images of the $La(OH)_3$ product, indicating the nanobelts width of tens to more than one hundred nanometers. Inset: Energy dispersive X-ray spectroscopy (EDS) analysis of the nanobelts, showing the presence of La and O (H cannot be detected by EDS). e,f) An arrow-shaped end (e) and cross section (f) of a nanobelt, exhibiting a thickness of 10 nm. [Reprinted with permission from Ref 188: Hu *et al.*, Adv. Mater., 2007, 19, 470–474. copyright 2007 Wiley-VCH Verlag Gmbh & Co. KGaA, Weinheim].

The simple hydrothermal method has also been used for straight growth of oxide nanofibers in liquid phase. After a hydrothermal treatment of $ZnCl_2$ solution with hydrazine hydrate at 150°C, uniform ZnO nanorods with width approximately 70 nm and length of 2 μm were obtained [191]. A number of 1D nanostructured ZnO with various morphology and arrays was synthesized by hydrothermal method with modified conditions [192-199]. Moreover, α- and β-MnO_2 single crystal nanowires were synthesized by a hydrothermal treatment of manganese sulfate aqueous solution with ammonium persulfate [200]. By means of proper control of the reactant concentration and temperature, this facile hydrothermal route can be extended to the preparation of other oxides such as TiO_2 [201-204], MoO_3 [205,206] and complex oxide [207] nanowires, nanotubes and, nanobelts. Surfactants have been used in hydrothermal process for synthesis of some kinds of hydroxides nanofibers. Nonionic Poly(ethylene oxide) (PEO) surfactant was used to grow boehmite nanofibers under hydrothermal conditions [208,209]. It was believed that the surfactant micelles play an important role in the nanofiber growth by directing the assembly of aluminum hydrate particles through hydrogen bonding with the hydroxyls on the surface of aluminum hydrate particles. Meanwhile a gradual improvement in the crystallinity of the fibers during growth is observed and attributed to the Ostwald ripening process. Nanofibers of boehmite were converted to γ-alumina by a subsequent thermal treatment, as elimination of H_2O between boehmite layers took place. Nevertheless, the conversion from boehmite to γ-alumina did not lead to substantial changes in 1D nanostructured morphology [210]. In addition, cationic surfactant, such as cetyltrimethylammonium bromide (CTAB), was used for synthesis of nanotubes of Al_2O_3 [211], nanofibers and nanobelts of MoO_3 [212], nanorods of ZnO [213] and nanofibers of mesoporous silica [214] in a solution based hydrothermal process. In addition, nanofibers of complex oxides with perovskite (ABO_3) or spinel (AB_2O_4) crystal structures have been synthesized by solution based hydrothermal methods in the presence of surfactants [215-217]. Core/shell bismuth telluride/bismuth sulfide nanorods with shell branching were fabricated by using a biomolecular surfactant, L-glutathionic acid (LGTA) [218].

Hydrothermal methods have also been applied for synthesis of metal nanofibers. Silver nanowires were synthesized by a facile hydrothermal process in the presence of glucose as reducing reagent. As shown in Figure 18, uniform silver nanowires with large aspect ratio were obtained by hydrothermal treatment of silver nitrate aqueous solution with glucose at 180 °C for 18 h [219]. The hydrothermal conditions may be varied when different reducing regents are used. By using gemini surfactant 1,3-bis(cetyldimethylammonium) propane dibromide as reducing regent, silver nanowires could be formed at a relatively low temperature of 100 °C. [220]. Ultra-long and uniform copper nanowires with controllable diameters of 30–100 nm, length up to several millimeters (aspect ratio >105) and tunable crystallinity were obtained by hydrothermal processing the complex emulsion of copper (II) chloride and octadecylamine at 120–180 °C [221] or by complex-surfactant-assisted hydrothermal reduction approach [222]. The surfactant present in reaction system functioned as reagent for formation of emulsion of precursor and as a reducing agent for formation of metal nanowires. More metal, such as gold [223], germanium [224], cobalt [225] nanowires were synthesized by such a simple hydrothermal route.

Compared with solid VLS route, hydrothermal synthesis is performed at much lower temperature, which brings in low cost and attracts attention for large scale production. The nanofibers of hydroxyapatite (HAP) have been reported to be synthesized using hydrothermal route near or above 100 °C [226-228]. Recently, an even lower temperature crystallization method was reported [229]· Using stimulated body fluid (SBF) as reaction medium, nano-sized single-crystalline HAP fibers have been synthesized in SBF solution at 37 °C without adjusting the pH values during the whole reaction time. The synthesized HAP nanofibers have a length ranging from 60 nm to 100 nm and a uniform diameter of 3–5 nm. The HAP nanofibers could be an ideal structure as a specific reinforcement in the biomaterials for bone tissue implantation. Moreover, silver telluride (Ag_2Te) nanowires was reported to be synthesized by a room-temperature solution-phase route [230].

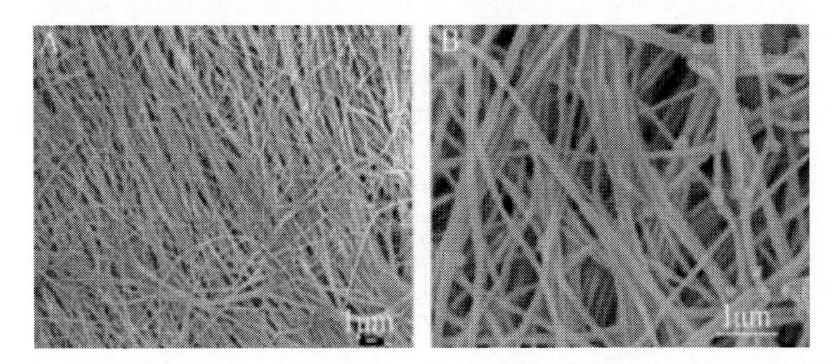

Figure 18. A) SEM image of the as-prepared silver nanowires. B) Image with higher magnification. [Reprinted with permission from Ref 219: Wang, *et al.* Chem. Eur. J., 2005, 11, 160 – 163. © 2005 Wiley-VCH Verlag GmbH&Co. KGaA, Weinheim].

1.5.2. Solvothermal Synthesis Route

Solvothermal synthesis route follows the same principle as hydrothermal methods and organic solvents are used as the media for dispersion of reactants.

Compound dissolved in solvents was brought to temperatures above the boiling point of mixture while the crystal growth was performed under autogenous pressure in a closed vessel. Similar to the hydrothermal process, ZnO nanofibers with large aspect ratios were fabricated in different organic solvents, such as ethanol and i-propanol[231-233]. Straight and uniform copper nanowires were facilely prepared via a solvothermal method by using copper nitrate as an inorganic precursor, and absolute ethanol served as a reducing agent as well as a solvent [234]. High-aspect-ratio TiO_2 nanofibers could be obtained at low temperatures and ambient pressure in nonpolar organic solvents [235]. A simple solvothermal approach was used to synthesize bimetal PbTe nanowires with tunable size [236]. The major advantage of solvothermal approach is that most materials can be made soluble in a proper solvent by heating and pressurizing the system close to its critical point.

Solvothermal process can be used to synthesize carbon nanotubes under a moderate condition [237]. Multiwall carbon nanotubes were synthesized by a low temperature solvothermal approach at 310 °C, in which ethoxylated alcohol polyoxyethylene ether was used as carbon source [238]. Transmission electron microscopy observations indicated that the obtained multiwall carbon nanotubes have outer diameters between 5 and 20 nm, and inner diameters between 2 and 8 nm. The length of the multiwall carbon nanotubes is of several microns. More reported carbon nanotubes [239,240], carbon nitride [241,242], and silicon nanowires [243] are fabricated by solvothermal process at much lower temperatures than that of solid based catalytic VLS growth route.

In addition, solvothermal method is a very important technology for fabricating 1D nanostructures of semiconductors at low temperatures [244,245], which is advantageous over hydrothermal method. CdS nanorods with varying dimensions were synthesized by solvothermal process [246]. It was observed that the anions present with the Cd-salts play an important role in determining the dimensions of the CdS nanorods. The crystalline nature of the sources was found to play a crucial role in determining the phase of the products. The nature of the sulfur source, molar ratio of the precursors, filling fraction of the solvent, and the synthesis temperature play important role in defining the size and shape of the products. By controlling the experimental parameters it was possible to control the dimension of the CdS nanorods within a certain range (diameter of the nanorods could be varied within a wide range from similar to 7 to 100 nm by varying the temperature within 100-250 ° C). Moreover, highly ordered large-area arrays of wurtzite CdS nanowires are synthesized on Cd-foil substrates via a simple liquid phase reaction route using thiosemicarbazide and Cd foil as the starting materials at 180°C [247].

Wurtzite CdS and CdSe nanostructures with complex morphologies such as urchin-like CdS nanoflowers, branched nanowires, and fractal nanotrees could be produced via a facile solvothermal approach in a mixed solution made of diethylenetriamine (DETA) and small amount of water (DIW) [248]. Urchin-like CdS nanoflowers made of CdS nanorods were in a form of highly ordered hierarchical structures, while the nanowires are branched nanowires, and the fractal CdS nanotrees were a buildup of branched nanopines. The results demonstrated that solvothermal reaction in a mixed amine/water can produce a variety of complex morphologies of semiconductor materials. The photocatalytic investigation indicated that the as-prepared branched CdS nanowires exhibited a high photocatalytic activity for degradation of acid fuchsine.

Moreover, InAs semiconducting nanowires with diameters of 7-70 nm and lengths of up to several microns were synthesized by a simple solvothermal method [249]. The preparation method features a low temperature (120-180 °C) and economical mass-production and is free of catalyst nanoparticles. This solvothermal route was a promising path towards the synthesis of other morphology-controllable one-dimensional (1D) III-V group nano-materials [250].

A large number of nanofibers of semiconductors have been fabricated by solvothermal method in recent years. Bi_2S_3 nanofibers have been synthesized by solvothermal route at a large scale and the application extended to V-VI (A_2B_3, A=As, Sb, Bi; B=S, Se, Te) group compound semiconductors [251-256]. More compound semiconductors, such as ZnS [257], CdE(E=S, Se, Te) [258-260], II-VI (ZnS, CdSe, ZnSe, CdS and CdO) [261,262] have been fabricated via solvothermal process. Compared with hydrothermal route, the disadvantage of solvothermal method is the disposal and cost of organic solvent and related environmental concern.

1.5.3. Sol-Gel Synthesis Route

Sol-gel process is a versatile solution based method for preparation of inorganic ceramic and glass materials. Under controlled conditions of sol-gel process, nanofibers of inorganic materials could also be obtained. The starting materials used in the preparation of the "sol" are usually inorganic metal salts or metal organic compounds such as metal alkoxides. In general, the sol-gel process involves the transition of a system from a liquid "sol" (mostly colloidal) into a solid "gel" phase. In a typical sol-gel process for synthesis of nanofiber materials, the precursor is subjected to a series of hydrolysis and polymerization reactions to form a colloidal suspension, the particles then condense along preferential 1D growth direction. Many kinds of nanofiber materials have been synthesized via simple sol-gel process, e.g. the fabrication of superconducting MgB_2 nanowires [263]. The solutions of $MgBr_2 \cdot 6H_2O$ and $NaBH_4$ were prepared by dissolving separately in ethanol under sonication. Surfactant of cetyltrimethylammonium bromide (CTAB) was added to each solution as directing reagent. The sodium borohydride solution was then added dropwise to the Mg-halide solution and the reaction mixture was sonicated for several minutes. This reaction mixture thickened over time and formed a gel when left open to the atmosphere for several hours. The thickened gel was loaded into a quartz boat, placed inside a horizontal tube furnace and the furnace temperature was ramped to 800 °C at ca. 10 °C min^{-1}. After thermal treatment for 5 min at 800 °C in an atmosphere of diborane and N_2, uniform nanowires of MgB_2 with uniform diameter were obtained as displayed in Figure 19. The formation of the gel in the initial reaction mixture could play the key role to the formation of the 1D nanostructured products with superconducting property. The successful bulk preparation of MgB_2 nanowires by simple sol–gel-assisted methods suggests that it may be possible to extrude and pyrolyze the precursor gel in a continuous process to produce long MgB_2 superconducting nanowire cables for practical applications.

Sol-gel method was also used for synthesis of 1D nanostructured metal and oxides [264,265]. A simple sol-gel process and subsequent calcination was employed to fabricate NiO nanowires [266]. Boehmite AlOOH nanofibers was synthesized by a modified sol-gel process with aluminum isopropoxide precursor at low room temperature. A subsequent thermal treatment at 600 °C resulted in γ-Al_2O_3 nanofibers with well preserved 1D nanostructure [267]. Moreover, sol-gel method is widely used for synthesis of complex oxides. A large number of complex oxides of perovskite (ABO_3) nanofibers were fabricated by modified sol-gel method. $(Na_{0.8}K_{0.2})_{(0.5)}Bi_{0.5}TiO_3$ (NKBT) nanowires were synthesized by sol-gel-hydrothermal process [268,269]. The morphology and structure analysis indicated that the sol-gel-hydrothermal route led to the formation of phase-pure perovskite NKBT nanowires with diameters of 50-80 nm and lengths of 1.5-2 μm, and the processing temperature was as low as 160 °C without the second step of further calcination. It was believed that the gel precursor and hydrothermal environment play an important role in the formation of the perovskite nanowires at a low temperature. Similar modified sol-gel-hydrothermal process was used for perovskite of $K_{0.5}Bi_{0.5}TiO_3$ [270] and $SrTiO_3$ [271].

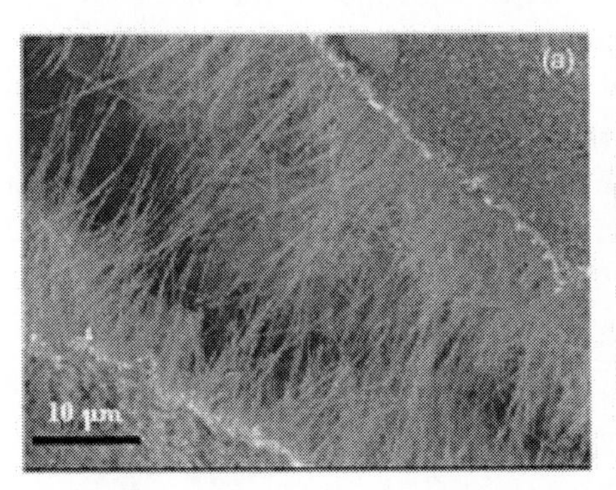

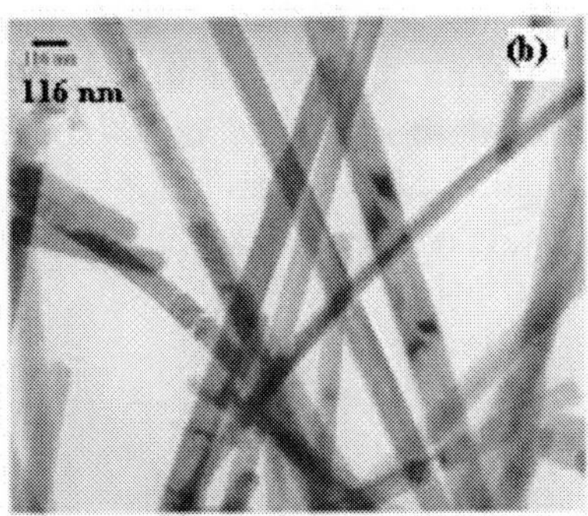

Figure 19. (a) SEM and (b)TEM images of MgB_2 nanowires. [Reprinted with permission from Ref 263: Nath *et al.*, Adv. Mater., 2006, 18, 1865–1868. © 2007 Wiley-VCH Verlag Gmbh & Co. KGaA, Weinheim].

Sol-gel process was combined with other techniques for synthesis of complex oxide nanofibers. By using AAO template, ZnO nanowires and nanotubes have been rationally fabricated within the nanochannels of porous anodic alumina templates by an improved sol-gel template process [272]. In this process, zinc nitrate and urea are used as precursors, where zinc nitrate serve as zinc ions source, and urea, offered a basic medium through its hydrolysis. ZnO nanowires or ZnO nanotubes can be obtained easily by controlling hydrolysis time. Similarly, rhombohedral structure p-type semiconductor Cr_2O_3 nanowires were generated by sol-gel template technology with the diameters in the range of 100-300 nm and the lengths ca. 10 μm [273]. Sol-gel route with nanoporous template (such as AAO) was applied for synthesis of nanofibers of complex oxides, such as $LaCoO_3$ [274], $LaFeO_3$ [275,276], $LaNiO_3$ [277], $La_{1-x}Ca_xMnO_3$ [278], $Bi_2Fe_4O_9$ [279] and $LiMn_2O_4$ [280].

Sol-gel process combined with electrospinning was employed to synthesis a variety of oxide ceramics [281]. As shown in Figure 20, $BaTiO_3$ nanofibers were synthesized via electrospinning combined with sol-gel process. The viscosity of mixed solution of barium acetate [$Ba(CH_3COO)_2$] and titanium isopropoxide [$Ti((CH_3)_2CHO)_4$] was controlled by poly(vinyl pyrrolidone) (PVP). Nanofibers with well-defined perovskite tetragonal phase were achieved by electrospinning at 20 KV and followed by thermal treatment at different temperatures. $BaTiO_3$ nanofibers with 50 nm in diameter and lengths up to 1 μm were found, which was a novelty in electrospinning of ferroelectrics [282]. More nanofibers of complex oxides, such as $ZnFe_2O_4$ [283] were fabricated by sol-gel process combined with electrospinning technique.

The sol-gel process allows to the synthesis of nanofibers materials at low temperature and the process could be scaled up for mass production. The formation of gel from solution is in the absence of metal catalysts and the resulting nanofibers have high purity and homogeneity.

1.6. Solid-Based Hydrothermal Synthesis

Solid-based hydrothermal synthesis (dry gel conversion) has been used for heterogeneous transformation of amorphous solid gel to crystallized zeolites [284] which deployed a dry process as opposed to the conventional solution based hydrothermal process. In the dry process, the hydrogel is dried and converted to crystals in the presence of water vapor (steam) or a mixture of vapors of water and organic structure directing agents. Compared with conventional solution based hydrothermal process, the solid based steam-assisted hydrothermal synthesis have advantages of simplicity in preparation, high efficiency of crystal growth, and can be operated in a large scale for mass production [285].

The solid based steam-assisted hydrothermal process was recently successful for synthesis of nanofiber materials. Boehmite nanowires with uniform diameters (12-16 nm) and length up to 1-2 μm was synthesized by a facile steam-assisted solid phase synthesis method [286]. Typically, aqueous solution of $Al(NO_3)_3 \cdot 9H_2O$ was precipitated by tetraethylammonium hydroxides (TEAOH) aqueous solution and control the desired pH value at 5.0±0.1. The resulting solid precipitates were recovered by filtration. The as-prepared solid cake-like wet gel was then transferred into a glass beaker (25 ml) sitting inside a Teflon vessel (200 ml), where 2 g of water were poured into bottom of vessel and physically separated from the solid wet-gel sample. The set-up is shown in Figure 21. When the autoclave was heated in oven, the small amount of water rose up as steam to assist the heterogeneous hydrothermal transformation of solid sample.

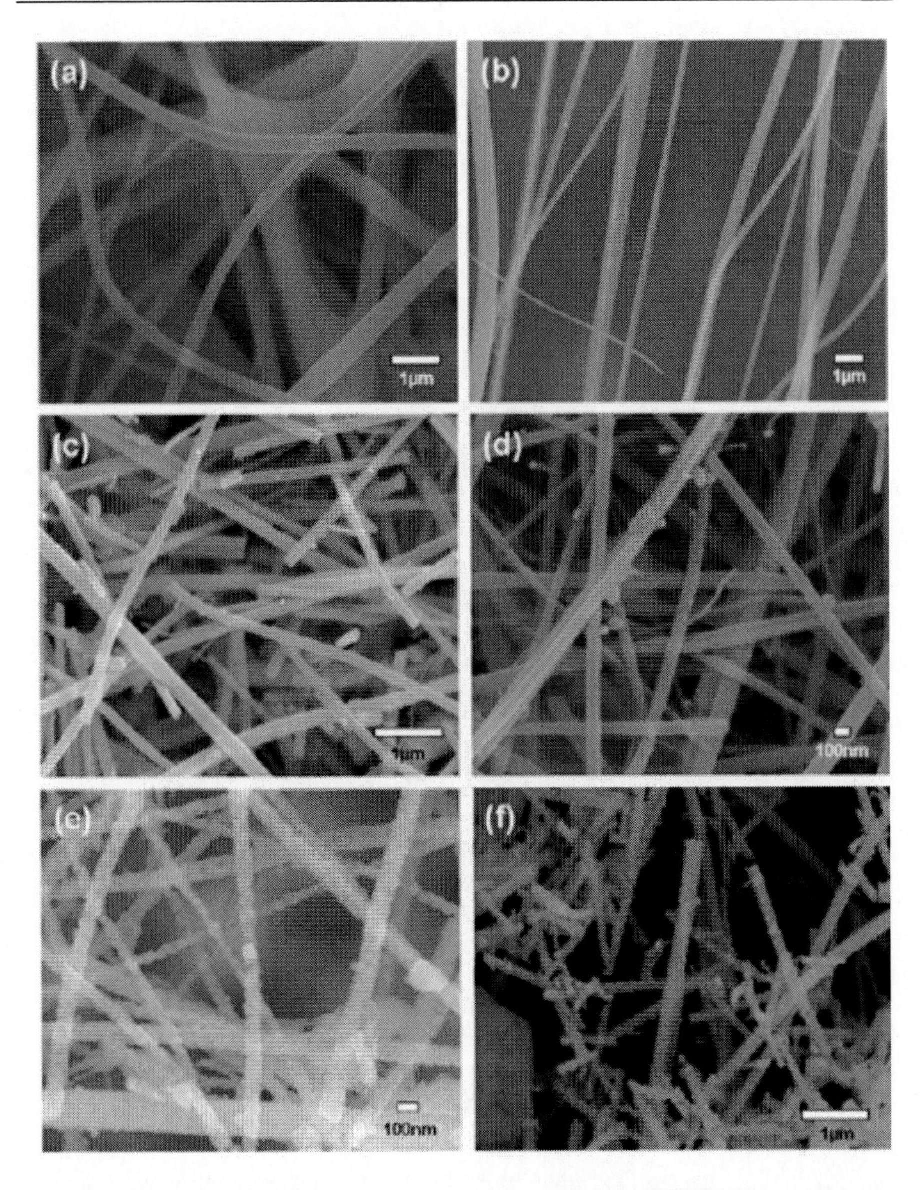

Figure 20. SEM images of $BaTiO_3$ nanofiber. (a) as-synthesized and dried at 120 °C, 1 h; (b) after polymer burnout at 450 °C, 3 h; (c) annealed at 580 °C, 24 h; (d) annealed at 580 °C, 48 h; (e) annealed 700 °C, 24 h; (f) annealed at 700 °C, 48 h. [Reprinted with permission from Ref 282: Yuh *et al.*, J. Sol-gel Sci. Tech., 2007, 42, 323-329, © Springer Science + Business Media, LLC 2007].

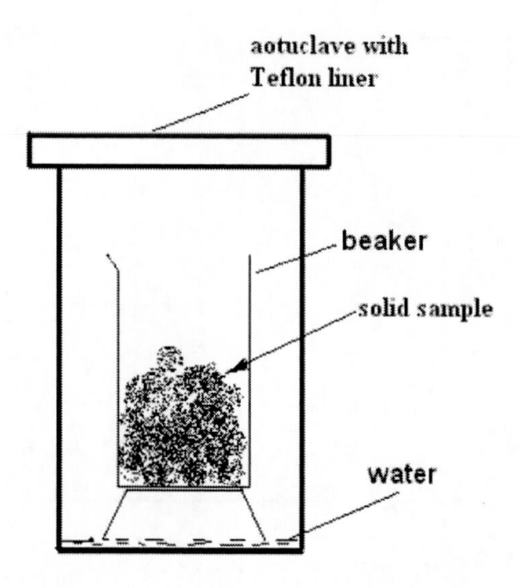

Figure 21. Scheme of set-up for solid based hydrothermal synthesis.

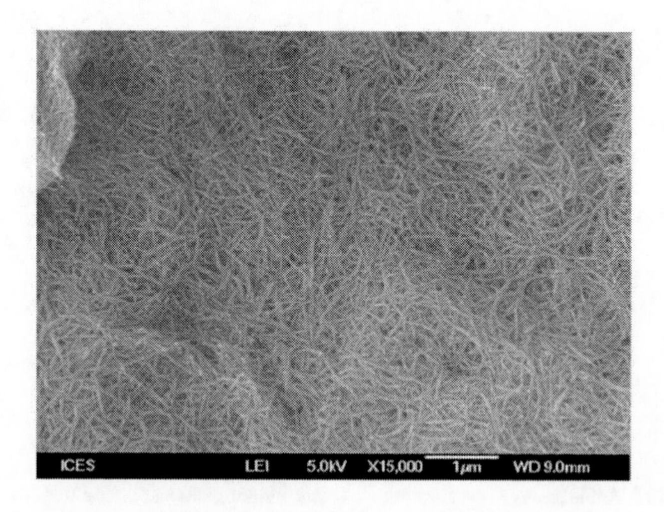

Figure 22. SEM images of boehmite AlOOH nanowires obtained from solid base steam-assisted hydrothermal synthesis.

After being sealed and heated at 170 °C for 72 h in the Teflon vessel, uniform nanowires of boehmite AlOOH was obtained. As exhibited in Figure 22, the nanowires of boehmite were clearly shown to be uniform in diameter at larger scale. This synthesis was performed by solid-based self-assembly process in the absence of organic solvent and crystal seeds. Popa and co-workers [287] proposed a mechanism of boehmite crystal formation in solution. The Al atoms in the boehmite lattice form a deformed octahedron with four oxygen atoms and two hydroxide neighbors. Such octahedra joined by edges result in AlO(OH) polymeric layers. These layers were held together by hydrogen bonds between the hydroxyl groups of each octahedron. The crystal growth in this synthesis was not performed in a solution; nevertheless, in the presence of steam, the local crystallization environment of the solid wet gel was an extremely supersaturated hydrothermal condition. Steam assisted the initial growth of boehmite and the extra water molecules formed during the crystallization of AlOOH maintained the suitable hydrothermal local condition for the rods to continuously grow. The crystallization of AlOOH has a preferential growth direction. Although the exact mechanism for the formation of the 1D nanostructure still remained unclear, the higher free energy of the amorphous aluminum hydroxides is believed to drive the crystal growth to Al(OH) in layered structure. The self-assembly fabrication of AlOOH nanowires from small amorphous precipitates was driven by a reduction of surface energy and the growth in 1D direction. The directed growth is limited by the stability of the planar boehmite layers, thus the length and diameter of the resulting nanowires are relatively uniform. The steam-assisted self-assembly fabrication of nanowires yielded high quality 1D nanostructures with clear-cut edge and high purity [286].

The structure transformation under steam-assisted crystallization and calcination at 600 °C was shown in Figure 23. Prior to steaming treatment, the precipitated solid resulting from reaction of $Al(NO_3)_3$ solution and TEAOH was in an amorphous state. No XRD peak is observed here for the dried powder of solid precipitate. After pressurized steaming treatment at 170 °C for 72 h, the amorphous phase was transformed to the crystalline phase. When the resultant solid was dried and ground to powder, a well-defined XRD pattern was observed and all diffraction peaks are perfectly indexed to boehmite AlOOH. The cell parameters of boehmite AlOOH are calculated to be: a=3.7093 Å, b=12.2365 Å, c=2.8685 Å, which are in good agreement with the values of bulk AlOOH (a=3.6936 Å, b=12.2141 Å, c=2.8679 Å, International Center for diffraction Data (ICDD), PDF file No 21-1307). No peak representing other phases was detected indicating the high purity of the resultant crystalline solids. After being calcined at 600 °C for 5 h, the boehmite AlOOH material was transformed to γ- Al_2O_3 through dehydration of internal water, as shown in Figure 23-c. The layered arranged boehmite AlOOH with hydrogen bond bridge was restructured to cubic-type close packing of oxygen atoms of γ-Al_2O_3 through internal condensation and removal of OH groups [288,289]. It was noted that the morphology of wire-like nanostructure was well preserved after being calcined at 600 °C for 5 h and converted to γ- Al_2O_3. The resulting nanowires of Al_2O_3 exhibited an excellent thermal stability by preventing sintering among the particles, compared with Al_2O_3 micro-powders. The high specific surface areas and the 1D nanostructure can be maintained up to 1300 °C, which offers high potential for applications in advanced ceramics and catalytic reaction at high temperatures.

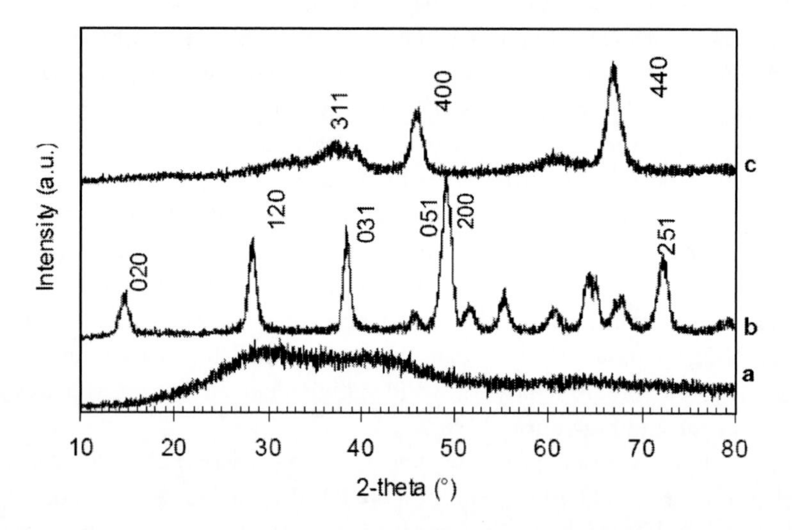

Figure 23. XRD patterns of (a) solid precipitates, (b) solid resulting from steam treatment of wet gel precipitate at 170 °C for 72 h, and (c) calcined sample-(b) at 600 °C for 5 h. [Reprinted with permission from Ref 286: Shen, *et al.*, J. Nanosci. Nanotechnol. 2007, 7, 2726-2733. © 2007 American Scientific Publishers].

The aspect ratios and morphology of boehmite nanofibers were found to depend on the regent and conditions for preparation of solid precipitates [290]. When $Al(NO_3)_3$ solution was precipitated with ammonia aqueous solution, straight nanorods of boehmite were obtained by the steam-assisted solid phase hydrothermal synthesis. In addition, the morphology of the crystallized nanostructures of boehmite AlOOH was found to be significantly affected by the precipitation condition used in the preparation of the precursor of wet gel solid. As indicated in Figure 24, when the wet gel precipitate was obtained at pH 5.0 using ammonia aqueous solution, straight nanorods of boehmite AlOOH with a diameter of 20-30 nm and length of 100 - 400 nm were observed (Figure.24-a). The aspect ratio of the nanorods boehmite AlOOH was up to 20. In particular, no irregular branching or bending was observed, which indicated that the AlOOH nanorods grow by self-assembly in 1D crystallization with high crystal perfection in the presence of steam. When the solid wet gel was obtained at neutral condition (pH7.0), nanorods of boehmite AlOOH was also observed after the same steaming treatment. The length of nanorods was shorter than that resulted from the wet gel precipitated at pH 5.0. The aspect ratio of nanorods boehmite AlOOH by steaming the wet gel precipitated at neutral condition was up to 10, although the diameter was unchanged. However, when the wet gel solid was obtained at alkaline condition, the morphology of resultant boehmite AlOOH was totally different from that of wet gel obtained under acidic condition. Nanoparticles of boehmite AlOOH of irregular shape were formed by steaming the wet gel solid precipitated at pH 10.0 (Figure 24-c) although the steaming conditions were the same as those for the nanorods displayed in Figure 24.a and Figure 24.b. This result indicated that the growth of nanorod in the presence of steam strongly depended on pH values at which the precipitation of amorphous aluminum hydroxides was carried out. An acidic precipitation condition was found to result in the solid wet gel with favorable environment for crystallization of AlOOH nanorods in 1D direction growth.

The steam-assisted hydrothermal treatment induced a transformation in chemical environment of aluminum species as investigated by MAS ^{27}Al NMR spectra. As shown in Figure 25, amorphous precipitates before hydrothermal treatment exhibited three types of aluminum species. The aluminum species of MAS ^{27}Al NMR were mainly represented by peak shift at 4.8 ppm, which was assigned to 6-coordinated Al species [291]. The additional two smaller peaks with chemical shift of 33.0 ppm and 63.4 ppm should be attributed to 5-coordinated and 4-coordinated Al species, respectively [292,293]. Upon wet gel steaming treatment, the aluminum species in the well crystallized nanorods of AlOOH had uniform chemical environment, thus, the resultant nanorods showed only one narrow ^{27}Al NMR peak corresponding to 6-coordinated Al species. After being calcined in open air at 600 °C for 5 h, AlOOH nanorods were converted to γ-Al_2O_3. Figure 25 (B) shows MAS ^{27}Al NMR spectrum of 1D nanostructured γ-Al_2O_3 nanorods. The main peak at 7.5 ppm for nanorods of γ-Al_2O_3 was attributed to 6-coordinated aluminum; and the smaller peak around 63.6 ppm, which was assigned to 4 coordinated aluminum species at tetrahedral sites [292]. The results confirmed the defective spinel structure of γ-Al_2O_3, where the amount of Al atoms at tetrahedral position is slightly less than half the amount of Al at octahedral sites [294,295].

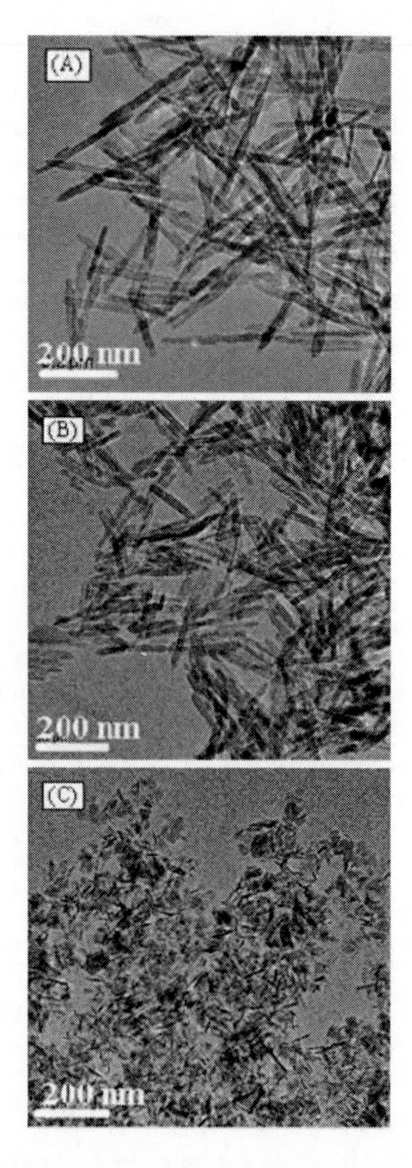

Figure 24. TEM images of crystallized boehmite AlOOH by steaming solid wet-gel precipitated under different conditions: (A) pH 5.0, (B) pH7.0 and (c) pH10.0. [Reprinted with permission from Ref 290 : Shen *et al.*, J. Phys. Chem. C, 2007, 111, 700-707. © 2007 American Chemical Society].

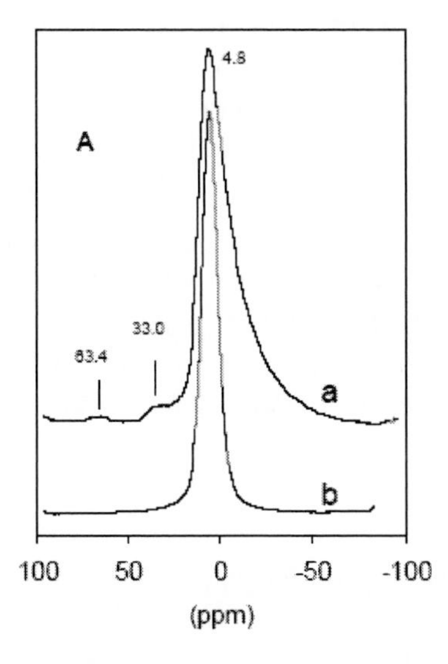

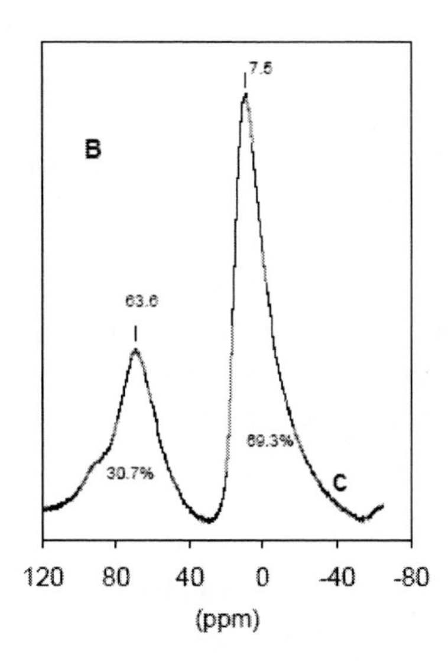

Figure 25. MAS ^{27}Al NMR spectra of (a) dried precipitate obtained at pH5.0, (b) boehmite nanorods obtained by steaming wet gel precipitate (pH5.0) at 200 °C for 48 h, (c) calcined sample-b at 600 °C for 5 h. [Reprinted with permission from Ref 290: Shen *et al.*, J. Phys. Chem. C, 2007, 111, 700-707. © 2007 American Chemical Society].

When the solid precipitate was dried and crashed to powder, a steam assisted dry-gel conversion process will transform the amorphous aluminum hydroxide precipitates to lace-like nanoribbons [296]. As shown in Figure 26, it was clearly seen that, after steam-assisted dry gel conversion at 200 °C for 48 h, lace-like nanoribbons materials were obtained. The image clearly shows the fibers in length of 1-2 μm and width of 100 nm. Figure 26-B exhibits the large area uniformity of the resulting nanoribbons indicated by the SEM image with low magnification. The image taken from lateral direction of nanoribbons with very high magnification reveals the thickness of the nanoribbons to be about 5-6 nm. No branching and impurity with irregular morphology are observed. As indicated in Figure 26-C, after calcination at 600 °C for 5 h, the morphology of the lace-like nanoribbons was well preserved.

This steam-assisted dry gel conversion method offers advantages in terms of process simplicity, avoiding the use of catalyst, surfactant, solvent or seeds while producing nanofibers of AlOOH boehmite and γ-Al_2O_3 with high purity and yield. This technique could potentially be further developed for fabrication of nanofibers of boehmite and γ-Al_2O_3 at large scale at low cost. In addition, this dry gel conversion method could be extended to the production of nanofibers of other oxide materials.

Furthermore, the solid-state reaction method was also reported for synthesis of other nanofiber materials, such as ZnO [297] $BaTiO_3$ [298] and polyaniline nanofibers [299].

The newly developed solid based hydrothermal method for fabrication of nanofiber materials is believed to be an efficient route. As the large-scale manufacturing of nanomaterials at an affordable cost stands the ultimate challenge for application of nanotechnology, this dry gel conversion potentially opens a route for transformation of laboratory nanotechnology to industrial manufacture.

1.7. Other Preparation Methods

In addition to the above described methods for fabrication of 1D nanostructured materials, a variety of methods have been investigated and applied of synthesis different materials.

1.7.1. Electospinning Method

Electrospinning method has been widely used for fabrication of polymer or polymer-inorganic composite nanofibers [300-302]. Different from "bottom-up" route by the controlled crystal growth of nanofibers through VLS, VS, hydrothermal, solvothermal and sol-gel process, electrospinning is a kind of feasible "top-down" process to generate nanofibers by using electrostatic forces to distort a pendant droplet of polymer solution into a fine fibers. Electrospinning is a simple and versatile process which could produce polymer nanofibers with diameters ranging from a few nanometers to several micrometers (usually between 50 and 500 nm) by using an electrostatically driven jet of polymer solution (or polymer melt). Various polymers have been successfully electrospun into ultrafine fibers in recent years mostly in solvent solution and some in melt form. For example, as displayed in Figure 27, conjugated polymer nanofibers of poly[2-methoxy-5(20ethylhexyloxy)-1,4-phenylenevinylene] (MEH-PPV) was effectively fabricated by electrospinning of the co-solution with poly(vinyl pyrrolidone) (PVP) [303]. The morphology and diameters of polymer nanofibers can be controlled by variation of electrospinning parameters.

A large number of polymer nanofibers[304-308] functionalized polymer nanofibers[309-313], polymer-inorganic hybrid nanofibers [314-318] were fabricated by the facile electrospinning methods.

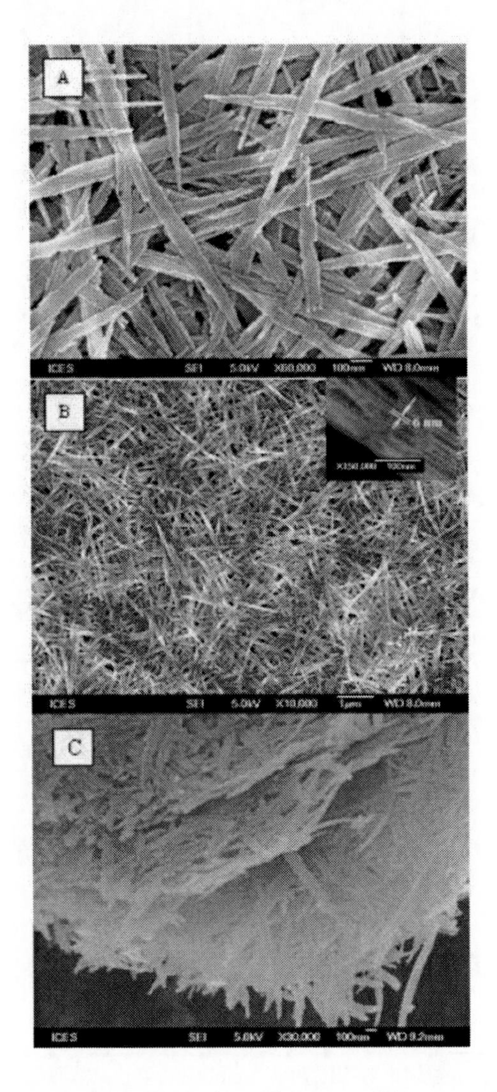

Figure 26. FESEM images of (A) lace-like nanoribbons synthesis by dry gel conversion at 200 °C for 48 h, (B) sample-A with low and high magnification (inset), and (C) γ-Al_2O_3 obtained by calcination of sample (A) at 600 °C for 5 h. [Reprinted with permission from Ref 296: Shen *et al.*, Mater. Lett., 2007, 61, 4280–4282. © 2007 Elsevier B.V.].

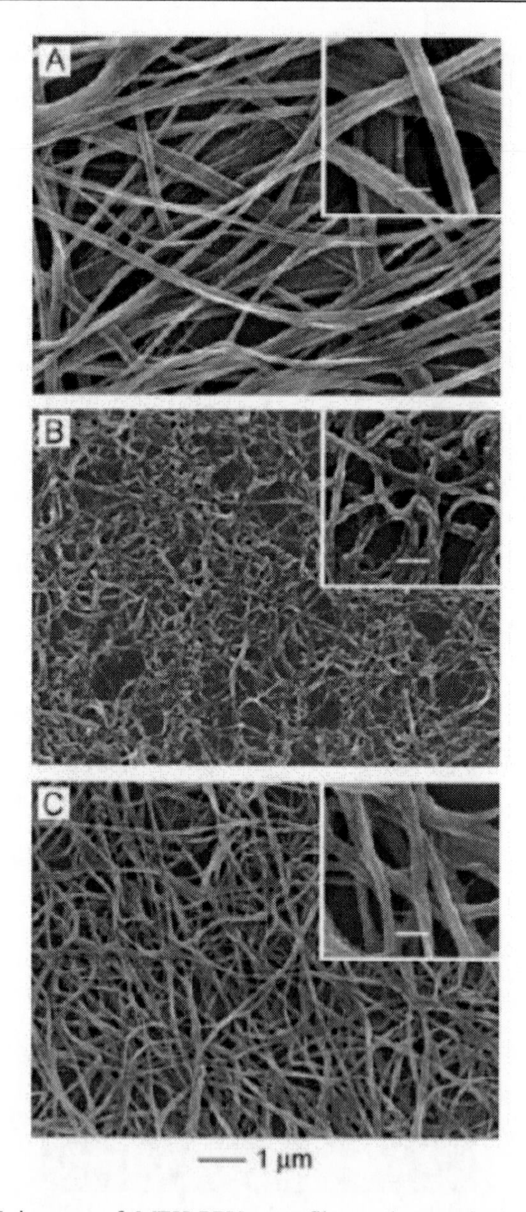

Figure 27. SEM images of MEH-PPV nanofibers electrospinning from solutions containing different concentrations of MEH-PPV and at different feed rates: A) MEH-PPV: 5 mg/ml, feed rate 0.3 ml/h; B) 1 mg/ml, 0.3 ml/h and C) 3mg/ml, 0.1 ml/h. The PVP component has been removed by ethanol extraction for all samples. The scale bar in the inset is 200 nm. [Reprinted with permission from Ref 303: Li *et al.*, Adv. Mater., 2004, 16, 2062- 2066. © 2004 Wiley-VCH Verlag Gmbh & Co. KGaA, Weinheim].

Moreover, the electrospinning method was also adopted for the synthesis of inorganic nanofibers, such as Al_2O_3 [319], SiO_2 [320], In_2O_3 [321], MoO_3 [322], V_2O_5-TiO_2 [323], TiO_2-Cr_2O_3 [324], $BaTiO_3$ [325], $CaMoO_4$ [326], La_2O_3-BiO-TiO_2 [327], GaN [328], BN [329] and metal nanowires [330]. Li and Xia [331] reported that hollow nanofibers with walls made of inorganic/polymer composites or ceramics could be prepared by electrospinning two immiscible liquids through a coaxial, two-capillary spinneret, followed by selective removal of the cores. Titania/polymer or anatase hollow nanofibers have been fabricated and the size and wall thickness could be independently varied by controlling a set of experimental parameters. The presence of a sol-gel precursor in the sheath liquid was necessary for the formation of stable, coaxial jets and hollow fibers with robust walls. The circular cross-section, uniform size, and well-controlled orientation of these long hollow nanofibers are particularly attractive for use in fabricating fluidic devices and optical waveguides.

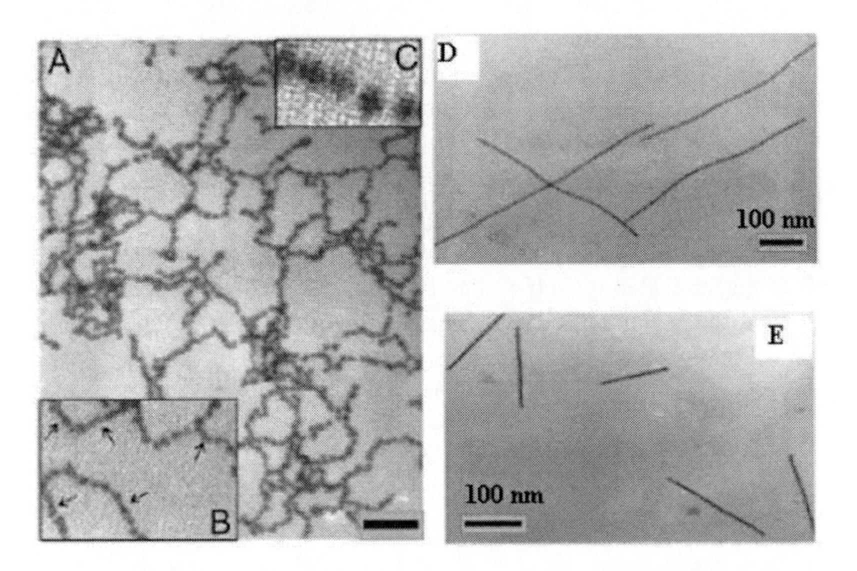

Figure 28. (A) TEM image of intermediate state of nanoparticle-nanowire transition for 5.4-nm nanoparticles. (B) The enlarged portion of the chain, with short rods marked by arrows. (C) The high-resolution TEM image of the adjacent nanoparticles in the chain. The "pearl necklace" aggregates were not observed in the standard dispersions of CdTe; And TEM images of CdTe nanowires made from 3.4- (D) and 5.4-nm (E) nanoparticles. Bars, 100 nm. [Reprinted with permission from Ref 337: Tang *et al.*, Science, 2002, 297, 237-240, © 2003 American Association for the Advancement of Science].

1.7.2. Self-Assembly

Spontaneous organization of nanoparticles to 1D nanostructures nanofiber attracted attention due to its convenience and high efficiency [332,333]. Single-crystal squaraine nanowires on silicon were synthesized by solvent evaporation-induced self-assembly [334]. Polypyrrole (PPy) micro/nanofibers were synthesized via a self-assembly process by using 4-hydroxy- 3-[(4-sulfo-1-naphthalenyl) azo]-1-naphthalenesulfonic acid (Acid Red B) as dopant and ferric chloride ($FeCl_3$) as oxidant [335]. It was believed that the micelles formed by the dopant and pyrrole monomer act as templates to form nanofibers during the assembly process. The nanobelts of peptide with fairly monodisperse widths on the order of 150 nm and lengths of up to 0.1 mm were also prepared via a self-assembly method [336]. Nanoparticles of CdTe were found to spontaneously reorganize into crystalline nanowires upon controlled removal of the protective shell of organic stabilizer [337]. As shown in Figure 28, the intermediate step in the nanowire formation was found to be pearl necklace aggregates. Strong dipole-dipole interaction is believed to be the driving force of nanoparticle self-organization. The linear aggregates subsequently recrystallized into nanowires and the diameter was determined by the diameter of the nanoparticles. The produced nanowires have high aspect ratio, uniformity, and optical activity. A paper published in 2005 reviewed the assembly of nanoparticles to 1D nanofiber through templated and template-free self-assembly routes [338]. Recently, more nanofiber materials were found to be fabricated by self-assembly of corresponding nanoparticles [339-344]. Nanofibers and tubes of polymers were synthesized through the oriented carbon–carbon cross-linking reactions towards rigid conjugated polymer networks and porous carbon nanofibers and nanotubes with high surface areas up to 900 m^2/g could be obtained after a thermal pyrolysis [345]. Magnetic field could also be deployed to induce the self-assembly of Fe_3O_4 nanoparticle to polyelectrolyte stabilized nanowires [346], and cobalt nanowires at low temperature [347].

Many other methods have been used for synthesis of 1D-nanostructured materials. Even for the same kind of nanofibers, various synthesis methods have been investigated. For instance, Zn(OH)F nanofibers were successfully synthesized by microwave irradiation [348]. Carbon nanofiber materials were synthesized by co-catalyst deoxidization reaction between $C_2H_5OC_2H_5$, Zn and Fe powder at 650°C [349]. Meanwhile, catalytic decomposition of hydrocarbon was also deployed for synthesis of carbon nanofibers at large-scale [350,351]. Plasma-enhanced vapor deposition process was developed for synthesis of carbon nanofibers [352,353]. The plasma-incorporated process was also applied for synthesis of metal oxides with large quantities at low temperature [354]. Polyaniline nanofibers were reported to be synthesized by using pseudo-high dilution technique in an aqueous solution [355] or by reverse microemulsion [356] as well as a solid phase mechanochemical synthesis route [357]. In addition, polymer nanofibers were also fabricated by centrifugal spinning [358]. Among different methods, each exhibited advantages and drawbacks in terms of quality of products, economical and environmental concern. Today, thousands of scientists are devoting the great effort to develop efficient process for fabrication of target nanofibers at low cost and at large scale for transformation of nanotechnology to industrial applications.

Chapter 2

PROPERTIES AND APPLICATIONS OF NANOFIBER MATERIALS

1D nanostructured fibers have unique anisotropic properties in terms of optical and electronic behaviors, which distinguished them from non-dimensional nanoparticles and bulk materials. The nanofibers of semiconductors and metal wires have been employed as connectors and switches to nanodevices, which are the crucial roles for the fabrication of nanodevices. It is also clear that 1D nanostructure with well-defined dimensions, composition, crystallinity and other chemical and mechanical distinct properties provide scientists a technology platform to explore a wide range of new applications such as nanodevices, sensor, reinforcement, energy storage and biomedical scaffolds.

2.1. NANODEVICES

The rapid progress in nanotechnology is fueling the design and manufacture of novel nanomaterials and fabrication of nano-scale devices. By manipulating the structure and composition of devices at nanometer/atomic scale, the performance can be greatly improved and new functionality can be developed. Due to the unique geometrical and physical property of 1D nanostructure, use of nanofibers (nanowires, nanorods and nanotubes) as nanocables has attracted much attention for connecting and controlling parts in nanodevices.

Nanofiber materials have demonstrated significant potential as basic building blocks for nanoelectronic and nanophotonic devices and also offer substantial promise for integrated nanosystems [359,360]. As shown in Figure 29, carbon nanotubes haven been used for fabrication of frequency performance field-effect transistor (FET) [361]. Metallic backgate carbon nanotubes FET were fabricated using standard microelectronic e-beam lithography combined with carbon nanotubes deposition by wet self-assembly. Highly resistive silicon substrate thermal oxide SiO_2 (260 nm) layer was grown prior to depositing coplanar access lines (titanium and gold). The metallic backgate was fabricated with aluminum. A thin gate dielectric layer of Al_2O_3 with thickness about 2 nm was obtained by O_2 plasma oxidation of aluminum. Carbon nanotubes were randomly deposited in the gate area to reach a density of about 10 CNs/μm as estimated by atomic force microscopy. Finally, palladium and gold (10 and 150 nm thick respectively) were deposited onto the carbon nanotubes to form the source and drain contacts. The source-to-drain spacing is 300 nm. This nanodevice of FET exhibited a current gain ($|H_{21}|^2$) cutoff frequency (f_t) of 8 GHz and a maximum stable gain value of 10 dB at 1 GHz, after de-embedding the access pads. It should be mentioned that alumina coating with thickness of 2 nm on nanofiber materials worked as important elements as gate dielectrics for fabrication of field-effect transistors as alumina exhibited a high dielectric constant (high-κ) and high bandgap of ~9eV, as well as its high thermal conductivity and stability [362].

Carbon nanotubes have also been investigated for the fabrication of FET in various ways and new properties were found [363-368]. The carbon nanotubes FET were found to have excellent on/off control ability at low supply voltage [369]. The high-performing single-tube field-effect transistor structures revealed that 1 to 1.5 nm was the optimum carbon nanotube diameter for high-speed digital applications, especially for the single-wall carbon nanotube fabricated nanodevice, quantum dot of single-wall-nanotube was an artificial atom with two- or four-electron shell structures [370]. Zeeman splitting of single particle levels was observed, which was advantageous for the spin based quantum computing device because the single spin was generated by putting one electron in the shell. Single-electron devices such as single-electron inverter and single-electron exclusive-OR gates have been fabricated to have advantages of small charging energy and the possible isolated device structure, which may be immune to the background-charge problem that is always a concern of the single-electron transistor nanodevices.

Semiconductor nanowires have been investigated for the fabrication of nanodevice of field-effect transistor [373-373]. Single ZnO and SnO_2 nanobelt was used for connecting the nano-enabled FET with excellent performance over conventional materials [374]. By controlling the backgate voltage, switch ratio as large as 6 orders of magnitude, a conductivity as high as 10 $(\Omega\ cm)^{-1}$ and a mobility as large as 35 cm^2/V was observed. Annealing SnO_2 nanobelt FETs in an oxygen-deficient atmosphere produced a negative shift in gate threshold voltage, indicating the generation of surface oxygen vacancies. This treatment provided an effective way of tuning the electrical performance of the nanobelt devices. The ability of SnO_2 FETs to act as gas sensors was also demonstrated to exhibit the high performance. In addition, ZnO nanobelt FETs were found to be sensitive to ultraviolet light [375].

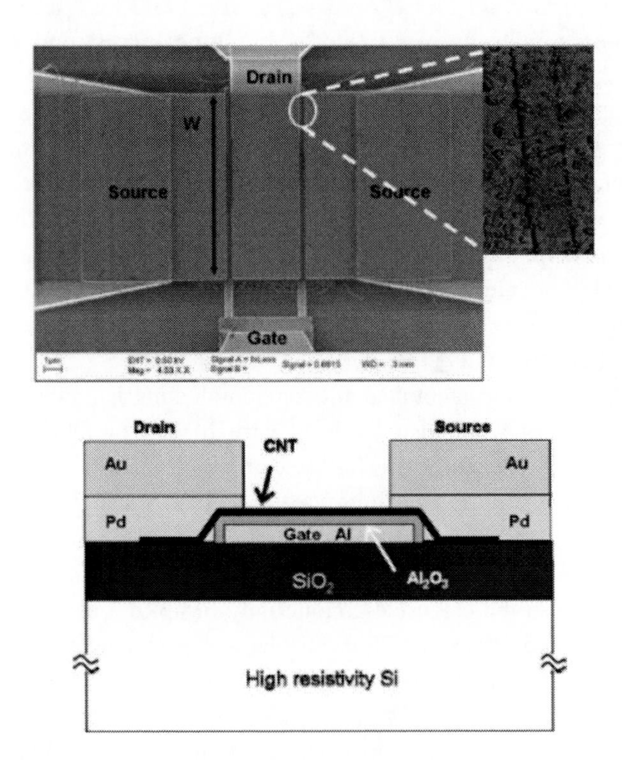

Figure 29. Scanning electron microscopy (SEM) image of the two-finger backgate carbon nanotubes FET active region. The inset shows an AFM image of a typical gate area where a random deposition of carbon nanotubes is visible with a density of about 10 tubes/μm. A scheme of structure of cross-sectional representation of the CN-based FET is shown [Reprinted with permission from Ref 361: Bethoux *et al.*, IEEE Electron Device Lett., 2006, 27, 681-683, Copy right IEEE].

Omega-shaped-gate (OSG) nanowire-based field effect transistors (FETs), as shown in Figure 30 have attracted a great deal of interest recently, because theoretical simulations predicted that they should have a higher device performance than nanowire-based FETs with other gate geometries [376]. OSG FETs with channels composed of ZnO nanowires were successfully fabricated by using photolithographic processes. In the OSG FETs fabricated on oxidized Si substrates, the channels composed of ZnO nanowires with diameters of about 110 nm were coated with Al_2O_3 using atomic layer deposition, which surrounds the channels and acts as a gate dielectric. About 80% of the surfaces of the nanowires coated with Al_2O_3 were covered with the gate metal to form OSG FETs. A representative OSG FET fabricated in this format exhibits a mobility of 30.2 $cm^2/(V\ s)$, a peak transconductance of 0.4 μS (V-g = -2.2 V), and an I-on/I-off ratio of 10^7 [376]. The value of the I-on/I-off ratio obtained from this OSG FET was higher than that of any of the previously reported semiconductor nanowire-based FETs. Its mobility, peak transconductance, and I-on/I-off ratio are remarkably enhanced by 3.5, 32, and 10^6 times, respectively, compared with a back-gate FET with the same ZnO nanowire channel as utilized in the OSG FET. The obtained enhancement of the electrical characteristics was mostly attributed to the OSG designed geometry and the passivation of the surface of the nanowire induced by the cladding of dielectric Al_2O_3.

Besides the carbon nanotubes and semiconducting nanowires, conducting polymer nanofibers were also used as core blocks in fabrication of FET nanodevices [377]. Recently, FET based on single nanowires of conducting polymers (i.e., polyaniline and polypyrrole) were fabricated and tested [378]. The 100-nm-wide and 2.5μm-long conducting polymer nanowire field effect transistors were turned “on” and “off” by electrical or chemical signals. A large modulation in the electrical conductivity of up to 3 orders of magnitude was demonstrated as a result of varying the electrochemical gate potential of these nanowires. Single nanowire conducting polymer field-effect transistors showed higher electrical performance than field-effect transistors based on conducting polymer nanowire electrode junctions and thin films in terms of their transconductance (gm) and on/off current (I_{on}/I_{off}) ratio. Furthermore, the performance of single nanowires conducting polymer field-effect transistors was found to be comparable to the silicon nanowire FET. These results imply that it is possible to tune the sensitivities of these conducting polymer nanowires by simple control of the electrolyte/liquid ion gate potentials. Furthermore, reversible switching was observed between conducting (ON) and nonconducting (OFF) states in the devices was reached by using monolayer of modified polyelectrolyte macromolecules fabricated in nanodevice, which showed a pronounced effect on the ON–OFF switching [379]. In addition, template synthesized conducting-polymer nanowires were found to produce electrically controlled nanoactuators operating in an aqueous electrolyte [380]. The artificial muscle nanowires exhibited unique electro-mechanical property, which enable controlled reversible expansion and contraction by electrical means. These nanoactuators are suitable for operation in a wide range of environment, such as in blood plasma or salt water, with potential applications in micro- and nanofluids for biomedicine, and environmental monitoring.

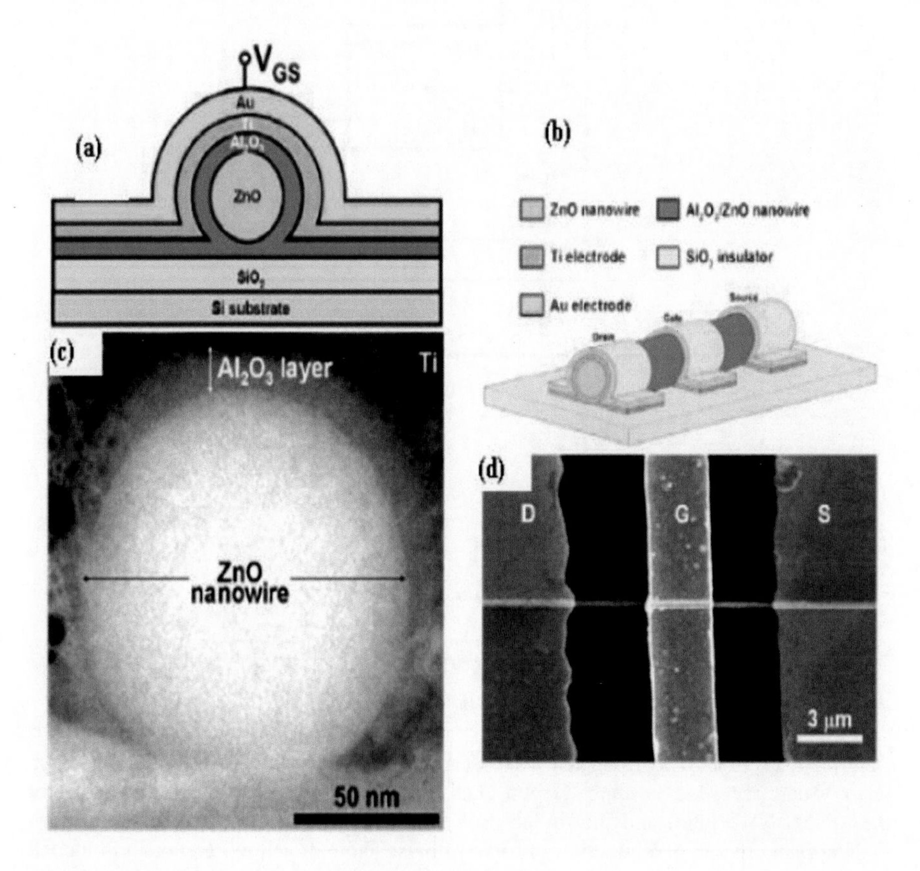

Figure 30. (a) Cross-sectional schematic, (b) Schematic top view , (c) TEM image and (d) SEM image of the OSG FET fabricated by the photolithographic process. [Reprinted with permission from Ref 376: Keem *et al.*, Nano Lett., 2006, 6, 1454-1458. © 2006 American Chemical Society].

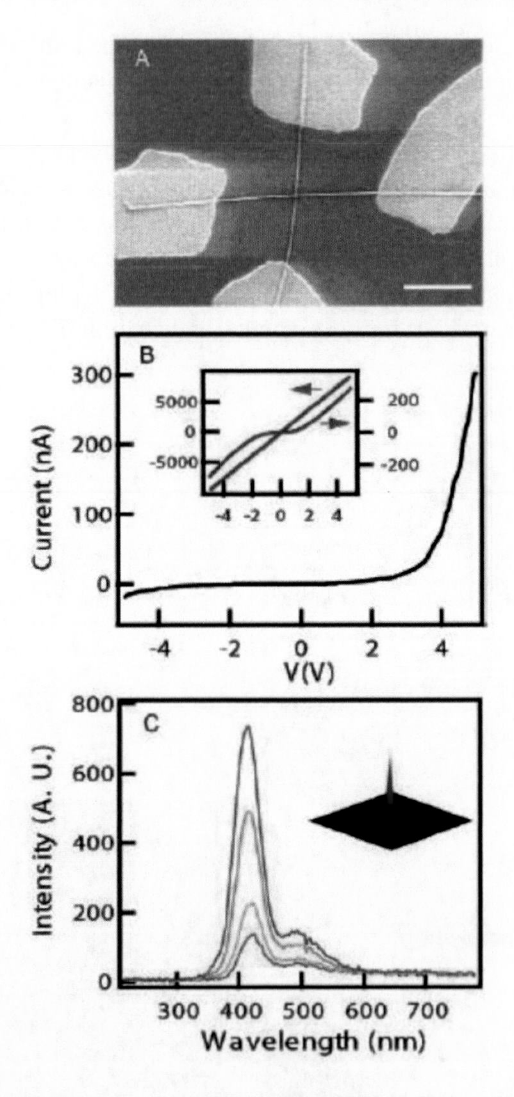

Figure 31. (A) FE-SEM image of a crossed GaN NW p-n device. Scale bar is 2 μm. (B) I-V data recorded for the p-GaN/n-GaN crossed NW junction. (inset) I-V data for the n-GaN NW (blue) and the p-GaN NW (red). (C) EL spectra recorded from a p-GaN/n-GaN crossed NW junction in forward bias. The red, yellow, green, and blue EL spectra were recorded with injection currents of 61, 132, 224, and 344 nA, respectively. (inset) Image of the emission from the crossed NW junction. [Reprinted with permission from Ref 381: Zhong *et al.*, Nano Lett., 2003, 3, 343-346. © 2003 American Chemical Society].

In addition to the widely investigated nanodevices of FET, potential nanoelectronics have been were fabricated by using semiconductor nanowires. New p-type GaN nanowire together with n- type GaN materials have been exploited to assemble complementary crossed nanowire p-n structures as shown in Figure 31 [381]. Transport measurements made on crossed nanowire p-n junctions show well-defined current rectification that is characteristic of p-n diodes (Figure 31B). Specifically, little current was observed in reverse bias to V_{sd} ~ -5 V, and there was a sharp current turn-on in forward bias at ca. 3.5 V. The *I-V* data recorded from the individual p-GaN and n-GaN nanowires were symmetric (inset, Figure 31B), and this rectification could be attributed to the crossed nanowire p-n junction but not to nanowire- metal contacts. In addition, the current turn-on at 3.5 V was consistent with the band-gap for GaN in this p-GaN/n-GaN device and contrasts the much lower current turn-on voltage for p-Si/ n-GaN heterojunctions. Crossed GaN nanowire structures assembled from p-and n-type materials showed that the nanoscale junctions behave as well-defined p-n diodes. Optical studies demonstrate a significant observation that these crossed GaN nanowire p-n junctions exhibit UV-blue light emission in forward bias. This enables p-n crossed nanowire junctions to function as nanoscale UV-blue light-emitting diodes (LED). Electroluminescence (EL) measurements on individual crossed GaN NW p-n diodes made using a homebuilt, far-field epifluorescence microscope showed that emission was localized at the nanowire cross point and was thus consistent with EL from the forward biased p-n nanojunctions (inset, Figure 31C). EL spectra recorded from forward biased nanoscale p-n junction exhibited a dominant emission peak centered at 415 nm and a smaller peak at ca. 493 nm (Figure 31C).

Recently, investigation into coupling the piezoelectric effect with the semiconducting property of ZnO nanowire has achieved a few unique potential applications [382,383]. It has been shown that an n-type ZnO nanowire can be used to produce a p–n junction that serves as a diode. As displayed in Figure 32, this design is based on the mechanical bending of a ZnO nanowire. In situ current–voltage (I–V) measurements and the manipulation of ZnO nanowires were carried out in a multiprobe nanoelectronics measurement (MPNEM) system. The two-terminal method was applied for electrical transport measurements at high vacuum to minimize influences from the environment. The potential energy barrier induced by piezoelectricity across the bent nanowire governed the electrical current transport through the nanowire. To quantify piezoelectricity, the I–V characteristics received at different levels of deformation were measured. The operation current ratio of a straight to a bent

ZnO nanowire could be as high as 9.3 : 1 at reverse bias. Thus, a mechanical force can be applied to control the electric output to determine the "1" and "0" states (ON and OFF states, respectively). Each nanowire array corresponds to a device element for memory application. With an applied –5 V bias, by applying a mechanical force, the operating current ratio range between a straight (-61.4 μA) and bent (-6.6 μA) ZnO nanowire. This leads to the appearance of well-defined "1" and "0" states; that is, the shape will be highly sensitive to the electric current due to piezoelectricity. A device element could be switched between these "1" and "0" states by mechanical force to receive different electric signals. On the basis of this switching mode, it is possible to characterize the elements as nanoscale electromechanical devices. This result implied that the semiconducting nanowire piezoelectric gate analogue diode has great potential to serve as a random access memory (RAM) unit. Moreover, the piezoelectric effect of nanowires may produce power output for nanodevice application. Nanowires of ZnO arrays grown on a flexible plastic substrate can serve as piezoelectric nanogenerators which can provide flexible power output as high as 50 mV [384]. This voltage generated from a single ZnO nanowire is enough to drive many nanoscale devices. Nevertheless, the mechanism of this ''energy harvesting'' from piezoelectric ZnO nanowires is still under debating [385,386].

More sophisticated nanodevices with advanced application potential have been investigated. As shown in Figure 33, Lieber and co-workers developed a new nanodevice model that information in the form of electrical impulses can fed into specific transistors in a nanocircuit, which is an essential step for carrying out computation [387,388]. Previous work had shown that where two nanowires cross function as a transistor. Applying an electric potential to an "input" wire will trigger a corresponding electric pulse in a perpendicular "output" wire. However, when a single input wire crosses several output wires, it can trigger multiple pulses simultaneously this is not suitable for computations. To overcome this drawback, a more selective nanocircuit was designed by Leiber and co-workers. They designed a grid of nanowires that initially consists of inactive transistors at each nanowire crossing. The wires at the specific or desired junction can be selectively activated by a designed chemical reaction. Traditional photolithography is used to direct a light-induced chemical reaction to specific crossbar intersections, which would steer electrical impulses to their desired destinations (as shown in Figure 33). In recent development, ZnO nanorods were integrated into sophisticated diode circuits for OR and AND logic operation or transistor circuits such as NOR and NOT logic gates due to high voltage and signal-power gains from the nanodevices [389]. These techniques offered great potential to connect nanosized circuits to large-scale integrated circuits, which can serve act as key technology to new generation of minisized computers with high performance.

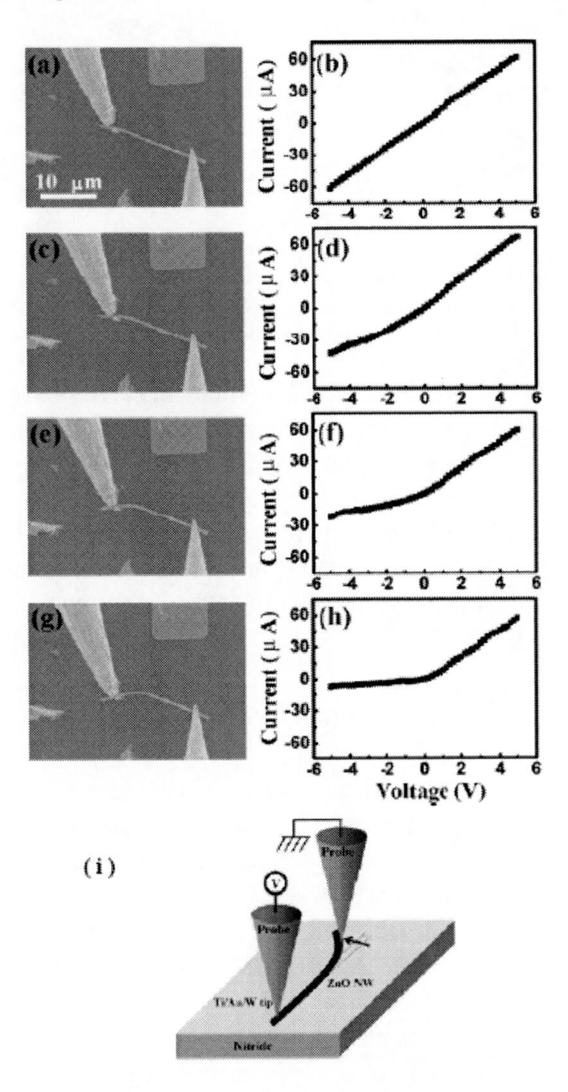

Figure 32. The sequence of SEM images of the ZnO nanowire at various bending angles is shown on the left (a, c, e, g). The corresponding *I–V* characteristics are shown on the right (b, d, f, h); (i) Schematic diagram of the nanomanipulation and in situ I–V measurement setup [Reprinted with permission from Ref 382 : He *et al.*, Adv. Mater., 2007, 19, 781–784, © 2007 Wiley-VCH Verlag Gmbh & Co. KGaA, Weinheim].

Figure 33. Chemical reactions at selected junctions control where current flows in a nanocircuit. [Reprinted with permission from Ref 387, Zhong, *et al.*, Science, 2003, 302, 1377-1379, © 2003 American Association for the Advancement of Science].

Nanodevices are not limited to nanofiber materials only, a number of nanodevices have been investigated using functional molecules [390,391], DNA [392], nanoparticles [393] and nanofilm [394]. Although the research in this nanodevice area is still at its infant stage, substantial progress and scientific advances have been achieved in the past few years. Nanoelectronics researchers are already beginning to bring basic nanoscience and nanotechnology research into our daily life.

2.2. Sensor Devices

Much interest have been also centered on using nanofibers in sensors for detecting of special chemical molecules, targeting the biomedical, environmental and hazard detection application. The large active surface areas of 1D nanostructured materials provided high sensitivity in terms of changing electrical conductivity towards different chemicals that are adsorbed on the surface. For instance, silicon nanowire fabricated into nanodevice exhibited ultrahigh sensitivity for label free detection of DNA [395]. When peptide nucleic acid (PNA) was captured in probe-functionalized silicon nanowire arrays, upon hybridization to complementary target DNA, the change in resistance is concentration-dependent and s linear over a large dynamic range with a detection limit of 10 fM. The silicon nanowire array biosensor discriminated satisfactorily against mismatched target DNA. It is also able to directly monitor the DNA hybridization event in situ and in real time. As shown in Figure 34, the resistance of the silicon nanowire strongly depended on the hybridization time for three different concentrations. The resistance increased upon exposure to the complementary DNA. The concentration of complementary DNA ranged from 1.0 fM to 1.0 nM and a noncomplementary DNA of 1.0 nM was used as control. The silicon nanowire arrays were immersed in the hybridization buffer, and its resistance was monitored while different amounts of DNA were added to the solution. While negligible resistance changes were observed at the control silicon nanowire array, the resistance was significantly changed after treated with the complementary DNA at different concentrations from 25 fM to 1.0 nM. The resistance of silicon nanowire array changed up to 280 % in 1.0 nM complementary DNA in TE buffer. The advantages of this biosensor are high uniformity and reproducibility, high yield, and excellent scalability and manufacturability. Compared to other DNA assays, the advantages of this silicon nanowire array biosensor are (i) ultrasensitive and label-free, (ii) more cost effective than optical biosensor arrays, (iii) rapid, direct, turbid, and light absorbing-tolerant detection, and (iv) portable, robust, low-cost, and easy-to-handle electrical components suitable for the development of gene expression profiling tools. A biosensor based on silicon nanowires modified by gold nanoparticles has also been fabricated to detect glutathione in solution via the cyclic voltammetry (CV) method [396]. Due to the strong sorption ability and high electrical conductivity of the modified silicon nanowires, the biosensor showed a fast response with good linear concentration dependence of glutathione in the range of 0.33 - 2.97 mu M. This facile modification approach could be extended to fabricate similar biosensors for the detection of various biomolecules.

More 1D nanostructured materials have been used for the detection of biomolecules [397-399]. Quantum-dot-modified nanotubes were also found to exhibit enhanced sensitivity to the detection of DNA hybridization [400]. Boron modified silicon nanowire sensor showed a wide linear range, high sensitivity, good reproducibility and long-term stability for the detection of glucose [401]. The integration of CdTe semiconducting nanowires and metallic nanoparticles in a single device leads to a biosensor with enhanced sensitivity that detects biomaterials over a broad range of concentrations [402]. Star and co-workers reviewed the application of carbon nanotubes based FET nanodevice serving as biosensor for the detection of proteins, antibody–antigen assays, DNA hybridization, and enzymatic reactions involving glucose [403]. Chitosan oligosaccharide derivative functionalized gold nanorods exhibited high sensitivity for protein detection and showed potential for applications such as biolabeling and biosensing [404].

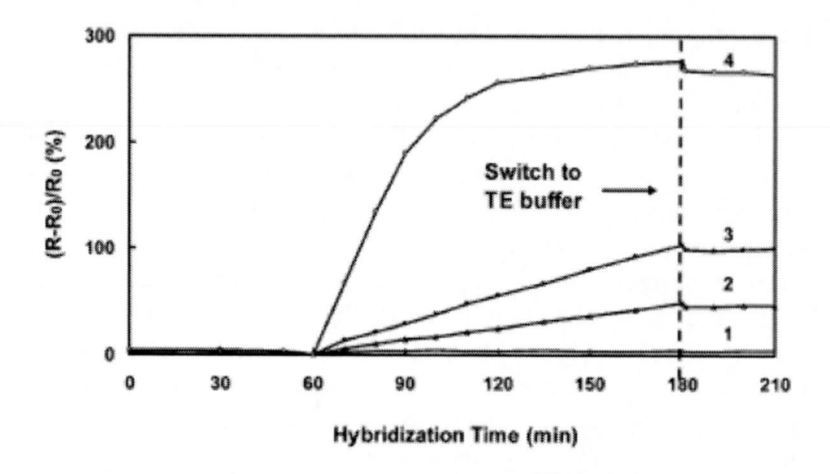

Figure 34. The dependence of resistance change of the silicon nanowire array biosensor on hybridization time in (1) 1.0 nM control, (2) 25 fM, (3)100 fM and (4) 1.0 nM target DNA in TE buffer. [Reprinted with permission from Ref 395: Gao *et al.*, Anal. Chem., 2007, 79, 3291-3297. © 2007 American Chemical Society].

Metal nanowires exhibited excellent performance as sensor for gas detection [405,406]. Palladium nanowires showed that the electrical resistance of Pd nanowire sensor decreases when hydrogen was adsorbed on Pd nanowires [407]. As displayed in Figure 35, the electrical resistance decreased with increasing concentration of hydrogen introduced. Once hydrogen was removed from gas stream the resistance restored to its initial level. The response and recovery times were quite fast at about 0.7 and 20 s, respectively and the sensing range of 0.2 ~1% hydrogen concentration was suitable for use in hydrogen safety sensors. The loss of electrical resistance is due to swelling of the nanowires as a result of hydrogen incorporation, which subsequent narrows the gaps between the nanowires. Sensors made of palladium nanowires also have a clear advantage of selectivity in detecting hydrogen concentration in the presence of other gases such as CO and CH_4 as compared with other types of hydrogen sensors. This advantage implies that the different sensing mechanism existing in the Pd nanowires as compared with the conventional Pd hydrogen sensors. Because of extraordinary high surface area of nanowires, the performance in terms of sensitivity and selectivity to detect hydrogen concentration was found to be by far superior to the conventional Pd sensors. A number of nanofiber materials have been investigated as sensors for the detection of hydrogen and ethanol [408-412]. Carbon nanotubes modified with palladium electrode also showed sensor response to medium concentration hydrogen gas (from 0.1% to 1.5% H_2) at room temperature [413]. Nanowires of ZnO, TiO_2 and $WO_{2.72}$ have been investigated to detect hydrogen, and the $WO_{2.72}$ nanowire (40 nm, diameter) exhibits the highest sensitivity (defined as the ratio of the sensor resistance in the gas concerned to that in air) of 22 for 1000 ppm at 298 K [414]. The $WO_{2.72}$ nanowire is also found to be good at sensing aliphatic hydrocarbons in the form of liquefied petroleum gas with a sensitivity of 15 for 1000 ppm at room temperature.

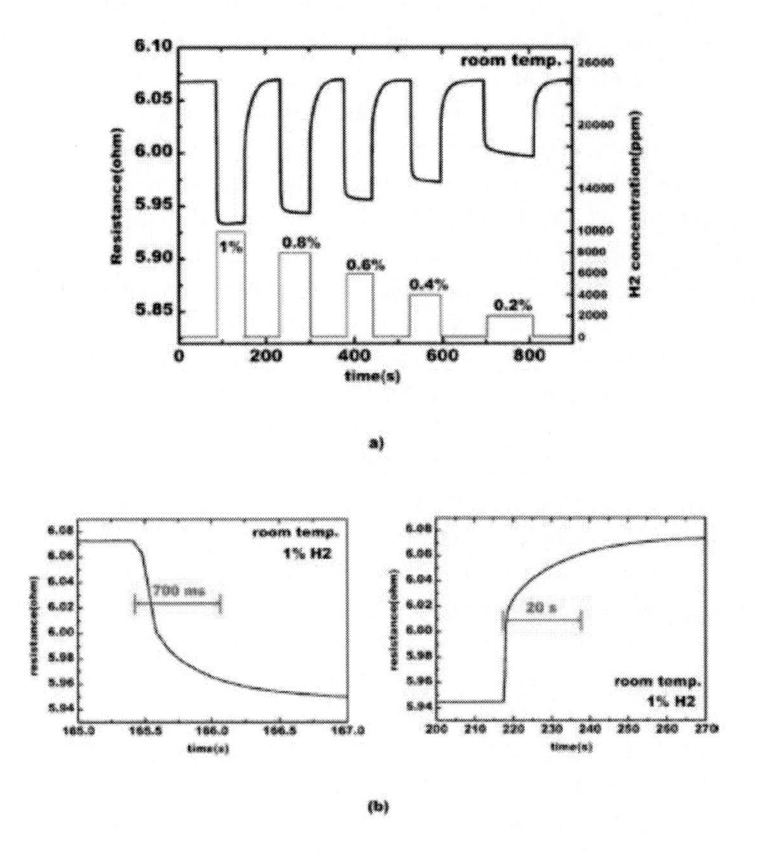

Figure 35. Electrical resistance measured as a function of time for the Pd nanowire hydrogen sensor. (a) Sensor resistance responses for hydrogen concentration varied in a range from 0.2 to 1% by pulses. (b) The fall and rise times (90% to signal saturation) measured at 1% hydrogen concentration. [Reprinted with permission from Ref 407: Kim *et al.*, IEEE Sensor J., 2006, 6, 509-513. © 2006 IEEE].

Nanofiber materials are also used as sensor for various other gases [415-417]. Recently, the nanorods of ZnO were synthesized via a hydrothermal process and fabricated as sensors. They showed high response and good selectivity to C_2H_5OH at 350 °C and high response to H_2S at room temperature [418]. In particular, the high sensitivity of nanostructured material fabricated sensors can detect 1 ppm C_2H_5OH and 0.05 ppm H_2S, which demonstrate clear advantages over the bulk oxides as sensors for gas detection. It was found that the ZnO-SnO_2 nanofiber film exhibited excellent ethanol sensing properties, such as high sensitivity, quick response and recovery, good reproducibility, and linearity in the range 3-500 ppm [419]. The sensing characteristics of ZnO, In_2O_3 and WO_3 nanowires have been investigated for the three nitrogen oxides, NO_2, NO and N_2O [420]. In_2O_3 nanowires with diameter of 20 nm prepared by using porous alumina membranes were found to have high sensitivity of about 60 for 10 ppm of all the three gases at a relatively low temperature of 150 °C. The response and recovery times were around 20 s. The sensitivity of these In_2O_3 nanowires was around 40 for 0.1 ppm of NO_2 and N_2O at 150 °C. WO_3 nanowires of 5 - 15 nm diameter prepared by the solvothermal process showed a sensitivity of 20 - 25 for 10 ppm of the three nitrogen oxides at 250 °C. The response and recovery times were 10 s and 60 s, respectively. The sensitivity was around 10 for 0.1 ppm of NO_2 at 250 °C. It was noted that the sensitivity of In_2O_3 and WO_3 nanowires was not affected by humidity condition even up to 90% relative humidity. Recent report showed that ordered silicon nanowire arrays on flexible plastics could push the detection limit of NO_2 down to low level of 10 ppb [421]. Ultralong Ag_2S nanowires were found to be very sensitive to oxygen and there was a quasi-linear relation between the current and the logarithm of oxygen pressure [422]. On the other hand, the individual Ag_2S nanowire exhibited remarkable photoelectric response to UV light. Illumination with 254nm UV light could make the current increase by as much as two orders of magnitude. These excellent performances indicate that Ag_2S nanowires are promising candidates for photo-switches and room-temperature oxygen sensors.

A variety of 1D nanostructured materials were developed as sensors with excellent performance to selectively detect target components. $WO_{2.72}$ nanowires showed high sensitivity to 2000 ppm of hydrocarbon (liquefied petroleum gas) at 200°C as well as relatively short recovery and response times [423]. Recently, relative humidity was measured by using CeO_2 nanowires. Both the response and recovery time were about 3 s, and were independent of the relative humidity. The sensitivity increased gradually as the humidity increased, and it was up to 85 at 97% RH [424]. The resistance decreases exponentially with increasing humidity, implying ion-type conductivity as the humidity sensing mechanism. In addition, TiO_2 nanowires were also found to exhibit high sensitivity for humidity [425,426]. Gold nanorods were found to have the outstanding selectivity and sensitivity for detecting mercury due to the well-known amalgamation process that occurs between mercury and gold [427]. This method presented a direct way to determine mercury in tap water samples at the parts-per-trillion level and showed excellent potential for monitoring ultra-low levels of mercury in water. Gold nanowires also exhibited transient optical response and showed an optical-switching effect as fast as 200 fs in the coupled spectrum, based mainly on the broadening and red-shift of the resonance spectrum of the plasmon polaritons, induced by electron–electron scattering and due to the strong optical excitation [428]. This feature revealed that the gold nanowires have high sensitivity of photonic device in response to slight disturbances of the particle plasmon resonance which is a very important characteristic for potential applications in sensors. In addition to sensing materials, carbon nanotubes films have showed the potential as strain/stress sensors by the dependence of Raman band shift on strain/stress [429]. Generally, nanofiber materials bring in superior performance as compared to counterpart bulk materials in terms of the excellent sensitivity and selectivity due to their large active surface areas and unique 1D nanostructures.

2.3. Drug Delivery and Tissue Engineering

Nanorods, nanotubes and nanowires attracted great attention in biomedical applications, such as drug and gene delivery [430,431]. Because of the nano-sized dimensions, the surface functionalized nanorods can immobilize drug molecules, protein or foreign genes for direct delivery to target cells to destroy malign cells or to provide supplement defective genes for "gene-therapy". Salem and co-workers [432] investigated the DNA delivery with bi-functional Au-Ni nanorods. Using molecular linkages that bind selectively to either gold or nickel, DNA and a cell-targeting protein were attached to the different segments. The surface of nickel segment was functionalized with 3-[(2-aminoethyl) dithio] propionic acid (AEDP) linker to bind DNA. On the other hand, transferrin, the first proteins exploited for receptor-mediated gene delivery, was bound to the gold segments of the nanorods through a thiolate linkage, by converting a small proportion of the primary amine groups of transferrin to sulphhydryl groups. A rhodamine tag on the transferrin provided a mechanism to confirm the presence of the nanorods in the cells and their intracellular distribution. The gene delivery potential was investigated by *in vitro* transfection experiments which were performed on the human embryonic kidney (HEK293) mammalian cell. The uptake of the DNA and protein attached nanorods were clearly shown by SEM and TEM investigation. Nanorods with compacted plasmids displayed a four-fold increase in GFP-positive cells in comparison with naked DNA. Luciferase expression by nanorod transfection was 255 times higher than that mediated by naked DNA. The results indicated that Au-Ni nanorods exhibited high potential in genetic vaccination approach with several advantages including ease of production and reduced risk of cytotoxicity and immunogenicity.

Gold nanorods have been intensively studied for cancer imaging and target drug delivery. [433-436] For effective delivery of gold nanorods through intravenous injection, it is important to develop gold nanorods with stealth character to avoiding excretion before uptake by cancer cells. Polyethyleneglycol (PEG) modified gold nanorods were found to reduce unspecific binding with blood components such as blood proteins and cells. From analysis of biodistribution of gold nanorods after intravenous injection, PEG-modified gold nanorods were stably circulating in blood with a half life of approximately 1 h, and there was no accumulation in major organs except for the liver at least for 72 h [437]. Due to the formation of a stealth character, PEG modified gold nanorods showed the potential for development of targeted delivery systems. When used in combinations with near infrared imaging and laser irradiating systems, simultaneous imaging and therapeutic can be carried out. Beside gold nanorods, porous iron nanorods [438] hollow magnetite nanorods [439] and silica nanorods [440] have also been investigated as nanocapsules for controlled drug delivery.

Among 1D nanostructured materials, biocompatible carbon nanotubes are good candidates for drug carriers [441,442]. Fictionalization of carbon nanotubes is a key step for the transformation of this new material into different systems for biomedical applications. Properly functionalized carbon nanotubes were examined to have a high propensity to cross cell membranes. The large surface area of carbon nanotubes can be charged with biologically active peptides, proteins, nucleic acid and drugs, which can then be delivered to the cell cytoplasm or organs. As carbon nanotubes exhibit low toxicity and are not immunogenic, they have emerged as novel alternative and efficient carrier for transporting of therapeutic molecules to desired destinations [443]. Recently, research of carbon nanotubes as drug carriers is very active for protein, DNA and drug delivery [444]. For effective drug delivery, multiwall carbon nanotubes have been covalently functionalized *via* 1,3-dipolar cycloaddition of azomethine ylides [445]. Two orthogonally protected amino groups on the sidewalls of CNT, subsequently derivatized with fluorescein isothiocyanate (FITC) and methotrexate (MTX) as indicated in Figure 36. FITC functions as fluorescent probes for bio-imaging. MTX is a drug widely used against cancer, but it suffers from low cellular uptake. By conjugation with carbon nanotubes, the limited cellular uptake could possibly to be overcome as the functionalized carbon nanotubes enhanced the penetration to cell membrane according to the results using cell of Human Jurkat T lymphocytes as receptors. The presence of the fluorescence probe on the tubes has allowed the analysis of its internalization and intracellular distribution. The multifunctional carbon nanotubes open the exciting opportunity for application in biomedical target delivery of anticancer drugs to tumor cells.

The single wall carbon nanotubes have been recently functionalized as "long boat" for platinum(IV) anticancer drug delivery. Moreover, surface functionalized carbon nanotubes have been used as carrier for DNA gene delivery [446]. Due to the hollow nanostructures and large surface area, carbon nanotubes exhibited effective delivery system to cells with high DNA uptake. Similarly, silica nanotubes were functionalized for delivery of DNA to target cells with improved efficiency [447].

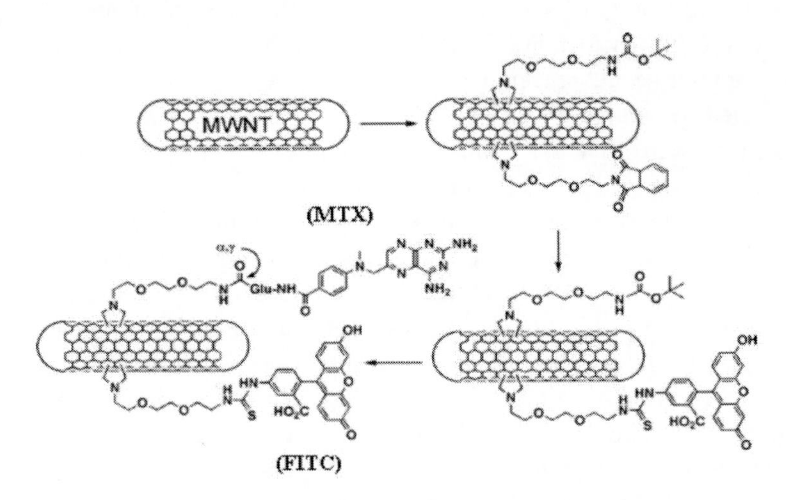

Figure 36. Scheme of route for bifunctionalization of carbon nanotubes as anticancer drug delivery and bio-imaging. [Reprinted with permission from Ref 445: Pastorin *et al.*, Chem. Commun., 2006, (11), 1182-1184. © The Royal Society of Chemistry 2006].

In addition to 1D nanostructures of nanorod and nanotube materials, polymeric nanofibers have been extensively investigated for diversified applications in biomedical and pharmaceutical fields [448]. Among the various currently used nanostructures for drug delivery applications polymeric nanofibers have received immense interest due to the ease of fabrication, controllable size/shape, and properties. Therapeutic agent loaded polymeric nanofibers could exhibit controlled/sustained release behavior for the delivery of these active agents for various therapeutic applications. As most the nanofibers of polymer materials have been fabricated by electrospinning at room temperature, the activity of drug was well preserved when the drug was loaded by mixing with polymer solution before electrospinning. The drug molecules were controlled released from the matrix of polymer nanofibers mainly by diffusional control. For instance, the release rate of heparin was well controlled by incorporation into electrospun poly(e-caprolactone) (PCL) nanofibers [449]. As shown in Figure 37, a total of approximately half of the encapsulated heparin was released from the heparin/PCL fibers after 14 days. A sustained release of heparin could be achieved from the fibers by the diffusional control during this period time. The fiber mats did not induce a pro-inflammatory response, and the released heparin retained biological functionality, indicating that the electrospun PCL fibers were a viable candidate as a drug delivery system for the localized administration of heparin to the site of vascular grafts. The local delivery of heparin to the site of vascular injury could prevent the myoproliferative response while avoiding the associated problems of systemic drug delivery. A variety of polymer nanofibers were used for controlled drug release [450-453]. Biodegradable poly(lactic-co-glycolic acid) (PLGA) nanofibers were used for sustained release of anticancer drug of paclitaxel [454]. An anticancer drug—paclitaxel-loaded PLGA micro- and nanofibers with diameters of around several tens nanometers to 10 μm were fabricated by electrospinning. The encapsulation efficiency for paclitaxel-loaded PLGA micro- and nanofibers was more than 90%. *In vitro* release profiles indicated that paclitaxel sustained release was achieved for more than 60 days. Thus, electrospun biodegradable micro- and nanofibers as implants could be an alternative approach toward the development of very promising local drug delivery devices for the treatment of brain tumor. Polymer nanofiber drug carriers were also found to facilitate the cellular uptake of an anticancer drug, lead to the efficient accumulation of anticancer drugs in drug-sensitive and drug-resistant cancer cells [455].

Polymer nanofibers have been used as matrix for protein controlled delivery and tissue growth factor [456,457]. Protein loaded with nanofibers of biodegradable polymer of poly(ε-caprolactone) (PCL) and poly(ethylene oxide) (PEO) could be fabricated by electrospinning. The obtained nanofibers of PCL/PEO blend meshes exhibited good morphological stability upon incubation in the buffer solution, resulting in controlled release of lysozyme over an extended period with reduced initial bursts [458]. By varying the PCL/PEO blending ratio, the release rate of lysozyme from the corresponding meshes could be readily modulated. The lysozyme release was found to be facilitated by increasing the amount of PEO, indicating that entrapped lysozyme was mainly released out by controlled dissolution of PEO from the blend meshes. This protein delivery strategy based on electrospun nano- and micro-fiber meshes can be potentially applied for various wound dressing and tissue engineering devices that require sustained release of angiogenic growth factors to the defect site.

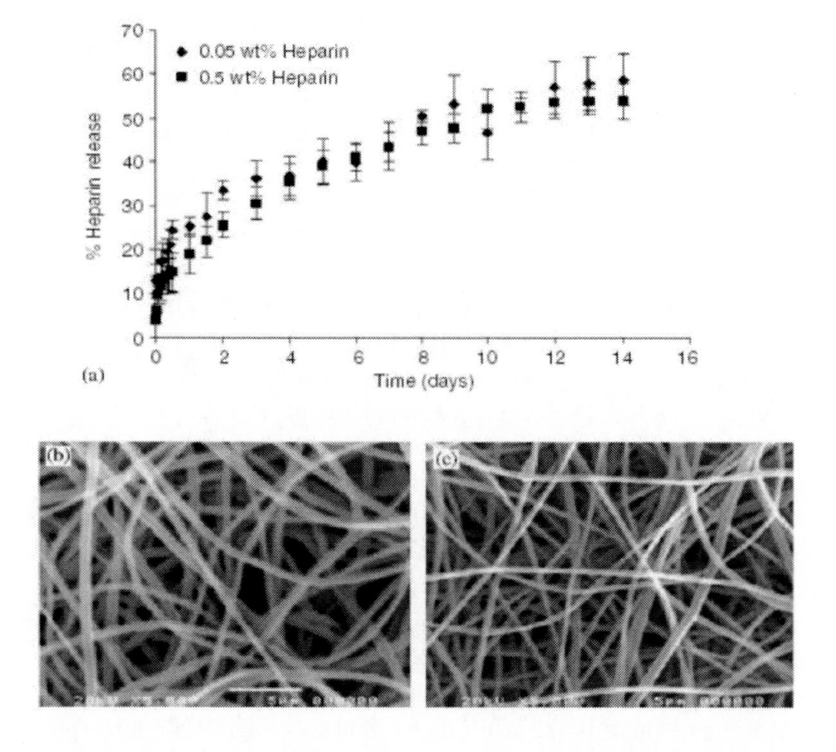

Figure 37. (a) Cumulative release of heparin from electrospun PCL fibers with different content of heparin at 37 °C, and SEM images of PCL fibers with (b) 0.05 and (c) 0.5 wt% heparin loadings. [Reprinted with permission from Ref 449: Luong-Van *et al.*, Biomater., 2006, 27, 2042–2050. © 2005 Elsevier Ltd.].

The nonwoven biodegradable polymeric nanofiber matrices are widely investigated as topical/local therapeutic agent delivery systems and as resorbable/biodegradable gauze for wound healing applications [459,460].

Nanofibers containing poly(vinyl pyrrolidone)–iodine complex (PVP–iodine) were obtained by electrospinning and it was found to be suitable for wound dressings [461]. The active iodine incorporated with nanofibers reacts with enzymes of the respiratory chains and with amino acids from the cell membrane proteins, resulting in destruction of the well-balanced protein tertiary structure and in irreversible damage to the microorganisms. This kind of functionalized nanofiber is a prospective material for antimicrobial wound dressing. Similarly, silver nanoparticles incorporated poly(vinyl alcohol) (PVA) nanofibers were fabricated by electrospinning of PVA/$AgNO_3$ solution and followed by UV radiation [462]. PVA/Ag fiber web was a good material for wound dressings because it had structural stability in moisture environment as well as quick and continuous release of the antimicrobial activity [463]. Furthermore, core-shell structured nanofibers have been developed and enabled independent control over the release of each drug by adjusting the physical and chemical properties of the colloids in the core, leading to a programmed release of multiple drugs [464].

The hydroxyapatite (HAp) nanoparticle incorporated biodegradable nanofibrous scaffold has been used for tissue regeneration [465-468]. Nanocomposites containing HAp nanoparticles have been shown to elicit active bone growth as HAp is a component of bone structure. Drug delivery and tissue engineering are actually closely related fields. In fact, tissue engineering can be viewed as a special case of drug delivery where the target is to achieve controlled delivery of mammalian cells. Controlled release of therapeutic factors in turn will enhance the efficacy of tissue engineering. Thus, biocompatible and biodegradable materials are needed for multifunctions as drug-delivery vehicles and tissue-engineering scaffolds. PLGA/HAp nanocomposites fit into this requirement. Fu *et al.* [469] investigated the bone regeneration under the controlled release of bone morphogenetic protein-2 (BMP-2) from PLGA/HAp nanofibrous scaffold. As BMP-2 is easy to get digested by enzyme once it is exposed to serum in vivo, sustained release provides the best strategy to maintain high level of BMP-2 in the local area, which is the main motivation to design the release profile of scaffolds. The author investigated two methods to load BMP-2 into nanofibrous scaffolds using an electrospinning method, including encapsulating into fibers or coating on fiber surface, and the results revealed that BMP-2 encapsulated into fibers retained its biological activity in vitro and in vivo. The addition of suitable amount of HAp nanoparticles can alter scaffold tensile strength and decrease residual solvent content. The PLGA/HAp composite scaffolds developed in this way exhibited good morphology/mechanical strength and HAp nanoparticles could be homogeneously dispersed inside PLGA matrix. Results from the animal experiments indicated that the bioactivity of BMP-2 released from the fibrous PLGA/HAp composite scaffolds was well maintained, which further improved the formation of new bone and the healing of segmental defects in vivo. As comparison, animal experiments demonstrated that BMP-2 loaded pure PLGA scaffold could not retain the bioactivity of BMP-2 in vivo and has no effect on bone healing. The results reflected the advantages of composite nanofibers over single component active ingredient carriers. The biocompatible and biodegradable HAp containing nanocomposites offer a platform for application in bone regeneration with the aid of controlled release of bone morphogenetic protein. On the hand, HAp/chitosan (Hap/CTS) nanofibers were fabricated by electrospinning and used for bone tissue engineering [470]. It was demonstrated that the incorporation of HAp nanoparticles into chitosan nanofibrous scaffolds led to significant bone formation oriented outcomes compared to that of the pure electrospun CTS scaffolds. The electrospun nanocomposites nanofibers of HAp/CTS, with compositional and structural features close to the natural mineralized nanofibril counterparts, are of potential interest for bone tissue engineering applications.

The 1D nanostructured material showed advantages in drug delivery applications as compared with traditional bulk materials in aspects of bioavailability, effective targeting, sustained release and potential cytotoxicity. Nanofiber materials promised a versatile nano-scale controlled or targeting drug delivery system based on nanorods, nanotubes and wire-like fibers. They can be used to deliver both small-molecule drugs and various classes of biomacromolecules, such as peptides, proteins, plasmid DNA and synthetic oligodeoxynucleotides to desired part of body. The composite nanofibrous scaffold also has potential applications for tissue regeneration while delivery bone morphogenetic protein-2. The diffusional control mechanism associated with nanofibers provided an effective sustained release of encapsulated drug molecules from nanofiber matrix.

2.4. Reinforcement of Composites

Nanofiber materials usually exhibit high specific mechanical strength than bulk counterpart materials. Among 1D nanostructures, carbon nanotubes and solid nanofibers are known to be the strongest and stiffest materials, either in terms of tensile strength or elastic modulus, which is superior to high-strength materials. In 2000, a multi-walled carbon nanotube was tested to have a tensile strength of 63 GPa [471] where as high-carbon steel has a tensile strength of approximately 1.2 GPa. Moreover, carbon nanotubes have very high elastic module, on the order of 1 TPa. Thus nanofiber materials have been explored as reinforcement element for thin surface coating, especially for polymer and biomaterials [472-474]. The nanofiber materials mixed with plastic polymers functioned similarly to "steel-in-cement" for construction of buildings with high strength. Multiwalled carbon nanotubes, covalently functionalized with chlorinated polypropylene, were used as the filler material in polymer-nanotube composites with improved Young's modulus and strength [475]. When 1vol% functionalized carbon nanotubes was loaded in the composite, close to a two-fold increase in both modulus and strength were achieved for polystyrene or poly(vinyl chloride) based composites. The high levels of reinforcement were due to the superior dispersion and stress transfer as evidenced by SEM images of the fracture surfaces. Furthermore, multiwalled carbon nanotubes were used as reinforcement for metal matrix [476]. Carbon nanotubes and Cu nanocomposites with homogeneously dispersed carbon nanotubes within Cu matrix could be fabricated by molecular-level mixing Cu ions with carbon nanotubes in a solvent. The yielded strength of carbon nanotubes reinforced Cu nanocomposites was three times higher than that of pure Cu. Carbon nanotubes have showed the most effective strengthening efficiencies among all reinforcement materials. The strengthening efficiency of reinforcement due to carbon nanotubes was eight times higher than SiC particles and eleven times higher than alumina fibers.

Nanocomposite fibers of single wall carbon nanotubes and Bombyx mori silk with increased Young's modulus were produced by the electrospinning process [477]· Regenerated silk fibroin dissolved in a dispersion of carbon nanotubes in formic acid was electrospun into nanofibers. The TEM investigation of the reinforced fibers shows that the single wall carbon nanotubes are embedded in the silk fibers. The mechanical properties of the single wall carbon nanotube reinforced fiber showed a significant increase in Young's modulus up to 460% in comparison with the un-reinforced aligned fiber. Similar reinforcement effect was observed for carbon nanotube and poly(methyl methacrylate) (PMMA) composite nanofiber fabricated by electrospinning [478]. Compared to pure PMMA nanofibers, Young's modulus and the tensile strength of the reinforced nanofibers exhibit considerable improvements due to the presence of carbon nanotubes in polymer nanofibers. Moreover, functionalized single wall carbon nanotubes were found to reinforce epoxy polymer composites through covalent integration [479]. Amino organic groups grafted on single wall carbon nanotubes converted it to be a very effective cross-linking agent. The functionalized nanotubes can be incorporated into epoxy composites through formation of strong covalent bonds in the course of epoxy curing reaction and became an integral structural component of the cross-linked epoxy composite. The single wall carbon nanotubes played a reinforcement role in the epoxy polymer matrix to dramatically enhance the mechanical properties. The ultimate strength and modulus were increased by 30-70% with the addition of only 1-4 wt.% of functionalized single wall carbon nanotubes. The nanotubes reinforced epoxy polymer composite also exhibited an increased strain to failure and high toughness. Moreover, for polyaniline fibers, small amount of carbon nanotube as reinforcement also significantly improved the electroactivity that translates to enhanced actuation performance [480]. The improved strength, stiffness and actuation of the composite fibers, which was named as high-strength artificial muscles, could find use in various applications where high-force operation is required, such as biomimetric musculoskeletal system.

Carbon nanofiber materials were found to function as solid lubricants and as reinforcement in thin surface coatings [481]. Reinforced polymer composite for surface coating significantly enhanced hardness and smoothness. The tensile strength of the nanocomposite film containing carbon nanofibers and ethylene/propylene (EP) copolymer was notably higher than that of the unfilled polymer although the crystallinity and the crystal orientation were not affected by nanofiber filler [482]. The thermal stability of the nanocomposite film was enhanced due to the reduced mobility of the polymer chains in the vicinity of polymer and nanofiber interface. It was also found that the carbon nanofiber as the reinforcement material effectively reduce the wear of polyphenylene sulfide (PPS) coating when sliding against the counterface of roughness 0.13–0.15 μm Ra [483]. The lowest steady state wear rate of 0.049 mm^3/km was obtained for PPS-10 vol.% carbon nanofiber composite. This wear rate was about one-seventh of that of the unreinforced PPS sliding against the same counterface. Metal matrix composite materials with suitable coating materials possess low coefficient of friction which can be used in a broad range of tribological applications, such as bearings, races, gears, splines, pins, hinges, and similar components used in aerospace based mechanical systems, aircraft propulsion and structures. In addition to carbon fibers, boehmite AlOOH nanorods were used as enforcement fillers to organic materials for hybrid surface coating [484]. Nanorods fillers of boehmite, with an aspect ratio around 20, significantly improve crack toughness of the glycidoxypropyltrimethoxysilane (GPTS) hybrid composite coating. The improvement was due first of all to the higher aspect ratio of the fillers, that tended to form a cross-ply structure which retards the growth of fractures; secondly the orientation of nanorods in the composite resulted in anisotropic toughness that was enhanced along the surface direction; finally chemical bonds were formed between boehmite nanofillers and GPTS during the sol–gel reaction [485].

1D nanostructured carbon materials possess excellent mechanical properties that can be utilized as reinforcement for biomaterials. Carbon nanofibers were used as reinforcement filler for improvement of mechanical properties of hydroxyapatite (HAP) composites [486]. The bending strength of carbon nanofiber reinforced HAP composites was similar to cortical bone. The fracture toughness values for carbon nanofibers containing HAP are around 1.6 times higher than those obtained for monolithic HAP ceramics. Carbon nanotubes were also used as reinforcement for imparting strength and toughness to brittle HAP bioceramic coating [487]. Plasma spraying assisted the uniform distribution of multiwalled carbon nanotubes reinforcement in HAP coating and it was found that the fracture toughness was improved by 56%, while the crystallinity was enhanced by 27% as compared with pure HAP. The reinforced HAP exhibited high biocompatibility by allowing unrestricted growth of human osteoblast hFOB 1.19 cells near carbon nanotubes, indicating that the uniform distributed carbon nanotubes promoted cell growth and proliferation.

The role as effective reinforcement materials is the unique properties of nanofibers with 1D nanostructure among different kinds of nanostructured materials. Non-dimension, nanoporous or 3D nanostructured materials might not perform the reinforcement function as excellent as that of nanofiber materials.

2.5. ENERGY STORAGE

With the increasing concerns over environment protection and the depletion on fossil fuels, hydrogen and the fuel cell technology have emerged as the best candidates for future energy sources. The challenge in using fuel cells with an on-board supply of hydrogen in automotive applications is to ensure that the hydrogen storage system is safe as well as light enough [488]. The development of effective hydrogen storage system is very active in recent years. Advance in nanotechnology offers great opportunities to improve the performance of existing energy storage systems. Applying nanoscale materials to energy storage offers higher capacity compared to the bulk counterparts due to the unique properties of nanomaterials such as large surface-to-volume atom ratio, and size-confinement effect. Among all potential nanomaterials for hydrogen storage, the 1D nanostructured carbon nanotubes and solid nanofibers are most promising candidates due to their low densities, high porosities and high specific surface areas suitable for hydrogen adsorption.

Carbon nanofibers are made up of very small graphite sheets that are stacked in specific configurations and separated by distances of 0.335 – 0.342 nm. Hydrogen has a kinetic diameter of 0.289 nm, which is slightly smaller than the 0.335–0.342 nm interlayer space of carbon nanofibers. When carbon nanofibers are placed in a vessel and exposed to hydrogen under pressures of 120-130 atm at room temperature, the hydrogen slips between the graphite sheets of the carbon nanofibers and adsorbs to surface of the carbon layers. The hydrogen storage capacity could be theoretically up to 40 wt% at hydrogen pressure of ~2000 psi if all the interstitials between the carbon sheets are filled with hydrogen. Nevertheless, the actual hydrogen adsorption on carbon nanofibers determined experimentally by most research groups is far below this value. A number of researchers reported the amount of hydrogen adsorption on carbon nanofibers at the level of 0.1~ 4 wt.% [489-492]. Few reports mentioned that hydrogen adsorption was measured at incredible high uptake up to 60 wt.% at 300 K and a pressure of about 100 atm on carbon nanofibers [493], but, different authors have failed to verify such a high adsorption on graphite nanofibers. Browning and co-workers observed a reasonable high level of hydrogen, up to 6.5 wt %, have been stored in carbon nanofibers under conditions of 12 MPa pressure and ambient temperature [494]. This capacity for hydrogen storage met the gravimetric target as set by the American Department of Energy (DOE). Unfortunately, due to the relatively low density of these nanostructures, typically 700 kg/m^3, they fall short of the desired volumetric storage requirement of 62 kg H_2/m^3 established by the DOE, delivering approximately 45 kg H_2/m^3.

Carbon nanotubes, a kind of hollow structured nanofiber, have been also intensively investigated as potential hydrogen storage materials. A large number of experiments and the work have been conducted on hydrogen storage using carbon nanotubes, and the several reviews were published before 2005 [495-497]. Hydrogen storage capacity below 2 wt% was usually achieved using either single-walled or multi-walled carbon nanotubes at 80 bar and room temperature [498]. The research in this area is still very active over the recent two years to improve the hydrogen adsorption on carbon nanotubes by modifying the surface with various methods [499]. Microwave plasma etching and Pd decoration methods were employed for the modification of carbon nanotubes for improved hydrogen uptake [500]. The defects on the nanotube wall resulted from etching and modification supplied more hydrogen accesses to the interlayers and hollow interiors of carbon nanotubes. The results of hydrogen uptake measurements showed that the etched carbon nanotubes had higher hydrogen storage capability than that of the original sample at ambient temperature and pressure of 10.728 MPa. Furthermore, the carbon nanotubes decorated with Pd showed a hydrogen storage capability of 4.5 wt.%, about three times higher than that of the non-decorated samples. To increase the hydrogen storage, the carbon nanotubes have been mediated by various methods, such as Ni-dispersion [501], benzene decoration [502], Pd doping [503], Pd-Ni-coating [504] KOH modification [505], Li doping [506], boron substitution [507]. Some hydrogen uptake results met the required value set by DOE. However, there still exist limitations on practical application. The hydrogen storage close to ambient condition remained as the main challenge subject for application of clean hydrogen energy in automotives [508]. In addition to carbon nanofibers, a variety of materials with 1D nanostructure have attracted attention for hydrogen storage [509]. Nanotubes of metal-diboride [510], boron nitride [511], carbon nitride [512,513], silica carbide [514] have been investigated as alternative materials for hydrogen storage. Boron nitride nanotubes were found to be advantageous over carbon nanotubes for hydrogen storage applications due to their heteropolar binding nature of their atoms. Moreover, porous carbon nanofibers also resulted in good electrochemical performance such as high reversible energy storage capacity and good cycle stability when they were used as anodes for rechargeable lithium-ion batteries [515]. Carbon nanotubes encapsulated nanofibers exhibited high performance for electrochemical energy storage, where a reversible capacity of 410 mA· h/ g could be achieved during 120 charge/discharge cycles when used as the anode in lithium based batteries [516].

2.6. Other Applications

Due to threadlike nanostructure, there is high potential in using nanofiber materials to fabricate membrane with mesh structure for filtration and separation application. Polymeric nanofibers with diameter less than 100nm have been synthesized by electrospinning process and the obtained fibers could be used for fabricating highly ordered membrane architectures [517-519]. The membrane made of nanofibers exhibited high filtration efficiency [520]. For instance, polyvinylidene fluoride nanofibers were electrospun into membranes and were successfully used for separation of 1, 5 and 10 μm polystyrene particles [521]. Membrane of surge sticks was fabricated by electrospinning and used for the separation of protein via molecular recognition [522]. Cellulose nanofiber membrane was prepared by electrospinning and surface was further functionalized with Cibacron Blue F3GA (CB), which is a general affinity dye ligand for the separation of many biomolecules [523]. However, due to limited chemical/thermal stability and swell of polymeric nanofibers, it is hard to achieve a high flux rate and the application is limited for separation in selected media. On the other hand, ceramic membrane fabricated with inorganic metal oxide nanofibers are of great interest in various separation process, owing to their chemical and thermal stability. Recently, high-performance ceramic membrane with a separation layer of inorganic nanofibers was developed [524]. The ceramic nanoporous filters were constructed on a porous substrate using larger titanate and smaller boehmite (AlOOH) nanofibers. The randomly oriented titanate nanofibers could completely cover the rough surface of the porous substrate of microsized α-alumina particles, leaving no pinholes or cracks. On top of this titanate fiber layer, a layer of γ-alumina fibers was formed using boehmite nanofibers. The resulting membranes can effectively filter out species larger than 60 nm at flow rates orders of magnitude greater than with conventional membranes, and they do not have the structural deficiencies of conventional ceramic membrane. A very thin membrane made of polymer-coated copper hydroxide nanostrands was prepared and has been used successfully for the separation of water-soluble proteins [525]. Functionalized moderately hydrophilic films with thickness of several tens of nanometers have also been fabricated, which exhibited a comparably high permeability to biological systems, and show potential for application in artificial kidney dialysis system.

The filtration of airborne particulate is essential for some environments, such as in clean room, equipment protection and air purification. Conventional filters made of microfibers with diameters in the order of 10 μm exhibit a local minimum in fractional collection efficiency for aerosol particle size range between 100 and 500 nm, which is called the most penetrating particle size [526]. The filtration efficiency depends strongly on the diameter of fibrous materials. High efficiency for airborne particulate filtration was achieved by carbon nanotubes coated on cellulose fiber filters [527]. With an extremely thin layer of multiwalled carbon nanotube coating, ultrafine particles (50-500 nm) filters can be collected at a very high efficiency >99%. This carbon nanotube-based filter could be used in aggressive chemical environments for capturing fine particulates, such as organic dyes, condensed lead fumes and even microorganisms such as viruses.

The use of 1D-nanostructured materials in various filtration devices is currently a promising area of further research in nanotechnology. These outstanding and compelling features of nanofiber materials are significantly superior for filtration applications as compared with other nanomaterials, such as nanoparticles with non-dimensional structure and nanoporous powder. The nanofiber membrane with excellent filtration efficiency is resulting in the rapid adoption by dairy, food, pharmaceutical, bioengineering, chemical, water treatment, and electronic industries for advanced value-added applications [528].

Apart from above-mentioned applications, there is a wealth of other potential of applications for 1D-nanostructured materials, such as in areas of catalysis [529-537], catalyst support [538-542], adsorbent for pollutant removal [543,544], solar collection [545-546], antibacterial [547] and all sorts. Special properties of nanofibers make them suitable for a wide range of emerging applications in years ahead, even though few commercial applications have yet to be realized as of today.

Chapter 3

RISK ASSESSMENT

Nanotechnology and nanoscience are expected to bring about a fundamental revolution in manufacturing in the coming decade and is receiving increased level of funding in global academic and industrial research. Similar to most matured technologies during the infancy period, the emphasis on benefits has been usually offset by considerable debate about the uses and safety of nanotechnology. Fortunately, along with existing and emerging use of nanoscale materials, growing concerns have arisen about their unintentional health and environmental impact. The toxicity of nanoparticles, especially in the format of airborne particles, has attracted much attention in getting better understanding of risks and to develop guideline on how to monitor and control worker exposure to these particles [548,549]. Risk impact studies of nanoparticles, such as TiO_2, SiO_2, ZnO, Fe_3O4, Al_2O_3, Cd-Te and CrO_3, on health have been assessed by in-vitro and in-vivo investigation [550-552]. Results of in vitro pulmonary cytotoxicity studies demonstrated a variety of responses to the different particle types, primarily at high doses to rat [553]. A recently published review/commentary on safe handling of nanoparticles has recommended strategic research strategies to support sustainable nanotechnology by maximizing benefits and minimizing environmental and health risks [554]. 1D nanostructured materials (nanorods, nanotubes, nanowires) also exhibited both technological advantages and adverse side-effects as health hazard as well as environmental concerns. Among 1D nanostructured materials, the toxicity of carbon nanotubes are under intensively research as large scale production has already been carried out and little information on potential health hazard arising from handling of nanomaterials was available [555,556]. Like nanoparticles, nanofiber materials could easily enter body either accidentally or deliberately introduced for drug delivery and bioimaging. It raises concerns on health hazard, especially on respiratory toxicity of nanofibers. High dose instillation exposed (5mg/kg) to single-wall carbon nanotube resulted in ~15% of rats within 24 hr post-installation [557]. The lethal effect was due to mechanical blockage of the large airways by instillation. Comparing among multiwalled carbon nanotubes, carbon nanofibers, and carbon nanoparticles, it was found that these nanomaterials generally led to proliferation inhibition and cell death, while carbon nanotubes appeared to be less toxic than other two materials [558]. Moreover, cytotoxicity was enhanced when the surface of the particles was functionalized after an acid treatment.

Although carbon nanotubes exhibited low immediate acute toxicity, they are biopersistent once they reach the lung of rats, and stimulate lung cells to produce TNF-α and induce lung inflammation and fibrosis [559]. Two month after intratracheal administration (0.5, 2 or 5 mg) to Sprague–Dawley rats, carbon nanotubes were still present in the lung. Pulmonary lesions induced by carbon nanotubes were characterized by the formation of collagen-rich granulomas protruding in the bronchial lumen, in association with alveolitis in the surrounding tissues. These lesions were caused by the accumulation of large carbon nanotubes agglomerates in the airways. Moreover, carbon nanotubes were also found to be respiratory toxicant to aquatic lives [560]. These results imply that carbon nanotubes are potentially harmful to humans with delayed effect.

Cytotoxicity tests using A549 human lung cells revealed the single-wall carbon nanotubes have very low acute toxicity [561]. However, *in-vitro* investigation on human cells revealed that nanofibers and nanotubes stimulate the release of the pro-inflammatory cytokine TNF-α and reactive oxygen species (ROS) from monocytic cells [562,563]. Cellular response varied with fiber morphology and state of aggregation; long, straight, well-dispersed nanowires produced significantly more TNF-α and ROS in monocytic cells. The monocytic cell phagocytic ability was reduced after exposure to all of the nanotubes. Brown and co-worker's microscopic examination of the cells after treatment with the nanotubes showed 'frustrated phagocytosis', implying that clearance of nanotubes from the human lungs by macrophages may be impaired [563].

Donaldson and coworkers published a review paper which discussed properties of carbon nanotubes and in relation to pulmonary toxicology and workplace safety [564]. The potential hazard of nanotubes was addressed with respect to aspects of length with fiber issue, surface areas, metal contamination, organic, aggregation, pathogenic processes and translocation of carbon nanotubes. Up to date, carbon nanotubes have been shown to have the potential in causing severe inflammatory and fibrotic reactions if they reach the lung of animals, these data still suggest the risk of developing serious lung diseases if workers were exposed to respirable carbon nanotubes for prolong period. It is clear that the understanding of the potential toxicity of carbon nanotubes is currently too scattered and that a number of gaps need to be filled in order to allow establishing sound science-based recommendations on guideline for monitoring and controlling exposure. Furthermore, besides the investigation on the respiratory toxicity of carbon nanotubes, information on hazard assessment of other 1D nanostructured materials are very limited.

The toxicity of nanostructured materials, including non-dimensional nanoparticle and 1D nanofibers, should be investigated using animals exposed to more realistic conditions, such as inhalation exposure, skin contamination or drug delivery dose, in order to verify the actual exposure risks in the workplace. When considering possible adverse health effects of inhaled particles or skin contamination, further studies should also address the possible carcinogenic potential of nanostructured materials. The goal of toxicological studies to identify the determinants of toxicity, in terms of dose, physical and chemical characteristics, susceptible populations, in order to make recommendations on safe handling during production and the use of the nanostructured materials below acceptable exposure levels. Strict industrial hygiene measures should to be adopted to limit exposure during manipulation of all nanostructured materials and not just nanotubes.

The toxicity research on nanomaterials is lagging behind the development of new nanostructures, but fortunately, the safety concerns are currently drawing more attention. It is believed that addressing safety concerns on nanostructured materials would not slow down the advancement of nanotechnology. Proper hazard control will ultimately benefit scientists and workers in implementation of nanotechnology to improve human life style.

Chapter 4

SUMMARY

Among various nanomaterials, 1D-nanofiber materials fall into the fast growing category due to their distinguished physical and chemical properties, which might not be achieved by other non-dimensional nanoparticles or the 2D nanoporous materials. The 1D-nanofiber materials function as core connecting parts in the fabrication of nanodevices due to the well-defined dimension, composition, and crystallinity with unique properties. They perform well in semiconducting, conducting and insulating elements at nano scale. Membrane fabricated using nanofiber materials exhibited excellent filtration properties for effective separation of biomolecules. In addition, the nanofibers also showed outstanding performance in sensing, drug delivery, and reinforcement in composite materials. Various current preparation methods have been summarized in this chapter. Semiconductor nanofibers are normally fabricated by solid-based VLS, VS growth process with the aid of other technologies, such as catalyst, high vacuum, CVD and plasma techniques. Nanostructured template confined growth is applied for the synthesis of metal and oxides nanofiber as well as hollow fiber nanotubes with uniform cross-section over a length of several hundred micrometers. Electrospinning under high voltage is suitable for the fabrication of polymer or polymer composite nanofibers. These preparation methods are facing the challenge of high cost, environmental pollution and low production rate. As comparison, wet chemical processes, such as hydrothermal, sol-gel, and steam-assisted crystallization route, are deployed for preparation of metal or metal oxide nanofiber materials at lower cost with potential for production at large scale. However, wet chemical method may not be applicable for fabrication of all kinds of nanofibers. All synthesis processes summarized here exhibited both merits and weaknesses for practical applications and there is an ongoing effort to strike a balance between the advantages and the costs. The limited understanding on the potential health hazards and environmental impact of nanostructured materials is an urgent issue along with the rapid development and application of 1D nanostructured materials.

REFERENCES

[1] Xia, YN; Yang, PD; Sun, YG; Wu, YY; Mayers, B; gates, B; Yin, YD; Kim, F; Yan, HQ. One-dimensional nanostructures: synthesis, characterization, and application. *Adv. Mater.,* 2003, 15, 353-361.

[2] Cademartiri, L; Ozin GA. Ultrathin nanowires—a materials chemistry perspective. *Adv. Mater.*, 2009, 21, 1013–1020.

[3] Wagner, RS; Ellis, WC.Vapor-Liquid-Solid mechanism of single crystal growth, *Appl. Phys. Lett.,* 1964, 4, 89-90.

[4] Wacaser, BA; Dick, KA; Johansson, J; Borgström, MT; Deppert, K; Samuelson, L. Preferential interface nucleation: an expansion of the VLS growth mechanism for nanowires. *Adv. Mater.* 2009, 21, 153–165.

[5] Roper, SM; Davis, SH; Norris, SA; Golovin, AA; Voorhees, PW; Weiss, M. Steady growth of nanowires via the vapor-liquid-solid method. *J. Appl. Phys.,* 2007, 102, Art No 034304 (7pp).

[6] Morral, AFI; Arbiol, J; Prades, JD; Cirera, A; Morante, JR. Synthesis of silicon nanowires with wurtzite crystalline structure by using standard chemical vapor deposition. *Adv. Mater.,* 2007, 19, 1347- 1351.

[7] Schmidt, V; Wittemann, JV; Senz, S; Gösele, U. Silicon nanowires: a review on aspects of their growth and their electrical properties. *Adv. Mater.* 2009, 21, 2681–2702.

[8] Gao, PX. Substrate atomic-termination-induced anisotropic growth of ZnO nanowires/nanorods by VLS process. *J. Phys. Chem. B.*, 2004, 108, 7534-7537.

[9] Chen, YX; Stevenson, I; Pouy, R; Wang, LD; McIlroy, DN; Pounds, T; Norton, MG; Aston, DE. Mechanical elasticity of vapour-liquid-solid grown GaN nanowires. *Nanotechnol.*, 2007, 18, Art No 135708-(8pp).

[10] Shi, WS; Zheng, YF; Wang, N; Lee, CS; Lee, ST. A General Synthetic Route to III-V Compound Semiconductor Nanowires. *Adv. Mater.,* 2001, 13, 591-594.

[11] Nguyen, P; Ng, HT; Kong, J; Cassell, AM; Quinn, R; Li, J; Han, J; McNeil, M; Meyyappan, M. Epitaxial Directional Growth of Indium-Doped Tin Oxide Nanowire Arrays. *Nano Lett.,* 2003, 3, 925-928.

[12] Chen, YQ; Cui, XF; Zhang, K; Pan, DY; Zhang, SY; Wang, B; Hou, JG. Bulk-quantity synthesis and self-catalytic VLS growth of SnO_2 nanowires by low temperature evaporation. *Chem. Phys. Lett.,* 2003, 369, 16-20.

[13] Liang, CH; Meng, GW; Lei, Y; Phillipp F; Zhang, LD. Catalytic growth of semiconducting In_2O_3 nanofibers. *Adv. Mater.,* 2001, 13, 1330-1333.

[14] Qiu, YF; Liu, DF; Yang, JH; Yang, SH. Controlled Synthesis of bismuth oxide nanowires by an oxidative metal vapor transport deposition technique. *Adv. Mater.,* 2006, 18, 2604-2608.

[15] Morales, AM; Leiber, CM. A Laser Ablation Method for the Synthesis of Crystalline Semiconductor Nanowires. *Science,* 1998, 279, 208-211.

[16] Zhang, YF; Tang, YH; Wang, N; Yu, DP; Lee, CS; Bello, I; Lee, ST. Silicon nanowires prepared by laser ablation at high temperature. *Appl. Phys. Lett.,* 1998, 72, 1835-1837.

[17] Kolb, FM; Hofmeister, H; Scholz, R; Zacharias, M; Gosele, U; Ma, DD; Lee, ST. Analysis of silicon nanowires grown by combining SiO evaporation with the VLS mechanism. *J.Electrochem. Soc.,* 2004, 151, G472-G475.

[18] Hu, JQ; Bando, Y; Zhan, JH; Liu, ZW; Golberg, D; Ringer, SP. Single-crystalline, submicrometer-sized ZnSe tubes. *Adv. Mater.,* 2005, 17, 975-979.

[19] Sood, DK; Sekhar, PK; Bhansali, S. Ion implantation based selective synthesis of silica nanowires on silicon wafer. *Appl. Phys. Lett.,* 2006, 88, 143110-143113.

[20] Jiang, Z; Xie, T. Yuan, XY; Geng, BY; Wu, GS; Wang, GZ; Meng, GW; Zhang, LD. Ctalytic synthesis and photoluminescence of silicon oxide nanowires and nanotubes. *Appl. Phys. A.,* 2005, 81, 477-479.

[21] Ye, CY; L.D. Zhang, X.S. Fang, Y.H. Wang, P. Yan, Zhao, JW. Hierarchical structure: silicon nanowires standing on silica microwires. *Adv. Mater.,* 2004, 16, 1019-1023.

[22] Nguyen, P; Ng, HT; Meyyappan, M. Catalyst metal selection for synthesis of inorganic nanowires. *Adv. Mate.,* 2005, 17, 1773-1777.

[23] Kamins, TI; Williams, RS; Basile, DP; Hesjedal, T; Harris, JS. Ti-catalyzed Si nanowires by chemical vapor deposition: Microscopy and growth mechanisms. *J. Appl. Phys.,* 2001, 89, 1008-1016.

[24] Westwater, J. Gosain, DP; Tomiya, S; Usui, S; Ruda, H. Growth of silicon nanowires via gold/silane vapor-liquid-solid reaction. *J. Vac. Sci. Technol. B.,* 1997, 15, 554-557.

[25] Sharma, S; Kamins, TI; Williams, RS. Synthesis of thin silicon nanowires using gold-catalyzed chemical vapor deposition. *Appl. Phys. A,* 2005, 80, 1225–1229.

[26] Morral, AFI; Arbiol, J; Prades, JD; Cirera, A; Morante, JR. Synthesis of Silicon Nanowires withWurtzite Crystalline Structure by Using Standard Chemical Vapor Deposition. *Adv. Mater.,* 2007, 19, 1347–1351.

[27] Dalacu, D; Kam, A; Austing, DG; Wu, XH; Lapointe, J; Aers, GC; Poole, PJ. Selective-area vapour-liquid-solid growth of InP nanowires. Nanotechnol. 2009, 20, Art. No. 395602(6 pp.).

[28] Yang, RB; Bachmann, J; Pippel, E; Berger, A; Woltersdorf, J; Gösele, U; Nielsch, K. Pulsed Vapor-Liquid-Solid growth of antimony selenide and antimony sulfide nanowires, *Adv. Mater.*, 2009, 21, 3170–3174.

[29] Chang, KW; Wu, JJ. Low-temperature growth of well-aligned β-Ga_2O_3 nanowires from a single-source organometallic precursor. *Adv. Mater.*, 2004, 16, 545-549.

[30] Wang, YW; Meng, GW; Zhang, LD; Liang, CH; Zhang, J. Catalytic Growth of Large-Scale Single-Crystal CdS Nanowires by Physical Evaporation and Their Photoluminescence. *Chem Mater.,* 2002, 14, 1773-1777.

[31] Hsieh, CT; Chen, JM; Lin HH; Shih, HC. Synthesis of well-ordered CuO nanofibers by a self-catalytic growth mechanism. *Appl. Phys. Lett.,* 2003, 82, 3316-3318.

[32] Duan, XF; Leiber, CM. General synthesis of compound semiconductor nanowires. *Adv. Mater.,* 2000, 12, 298-301.

[33] Morber, JR; Ding, Y; Haluska, MS; Li, Y; Liu, JP; Wang, ZL; Snyder, RL. PLD-Assisted VLS Growth of Aligned Ferrite Nanorods, Nanowires, and Nanobelts - Synthesis, and Properties. *J. Phys. Chem. B,* 2006, 110, 21672-21679.

[34] Colli, A; Hofmann S; Fasoli, A; Ferrari, AC; Ducati, C; Dunin-borkowski, RE; Robertson, J. Synthesis and optical properties of silicon nanowires grown by different methods. *Appl. Phys. A,* 2006, 85, 247–253.

[35] Persson, Ai; Larsson, MW; Stenström, S; Ohlsson, BJ; Samuelson, L; Wallengerg, LR. Solid-phase diffusion mechanism for GaAs nanowire growth. *Nature Mater.,* 2004, 3, 677-681.
[36] Dick, KS; Deppert, K; Karlsson, LS; Wallenberg, LR; Samuelson, L; Seifert, L. A new understanding of Au-assisted growth of III-V semiconductor nanowires. *Adv. Funct. Mater.,* 2005, 15, 1603-1610.
[37] Campos, LC; Tonezzer, M; Ferlauto, AS; Grillo, V; Magalháes-Paniago, R; Oliveira, S; Ladeira, L; Lacerda, RG. Vapor–Solid–Solid growth mechanism driven by epitaxial match between solid AuZn alloy catalyst particles and ZnO nanowires at low temperatures. *Adv. Mater.*, 2008, 20, 1499–1504.
[38] Gu, G; Burghard, M; Kim, GT; Düsberg, GS; Chiu, PW; Krstic, V; Roth, S; Han, WQ. Growth and electrical transport of germanium nanowires. *J. Appl. Phys.,* 2001, 90, 5747-5751.
[39] Yu, JY; Chung, SW. Heath, JR. Silicon nanowires: Preparation, device fabrication, and transport properties. *J. Phys. Chem. B,* 2000, 104, 11864-11870.
[40] Chung, SW; Yu, JY; Heath, JR. Silicon nanowire devices. *Appl. Phys. Lett.,* 2000, 76, 2068-2070.
[41] Wang, N; Tang, YH; Zhang, YF; Lee, CS; Lee, ST. Nucleation and growth of Si nanowires from silicon oxide. *Phys. Review B,* 1998, 58, R16024-16026.
[42] Zhang, RQ; Lifshitz Y; Lee, ST. Oxide-assisted growth of semiconducting nanowires. *Adv. Mater.,* 2003, 15, 635-639.
[43] Zhang, RQ; Zhao, MW; Lee, ST. Silicon monoxide clusters: the favorable precursors for forming silicon nanostructures, Phys. *Rev. Lett.,* 2004, 93, 0955031-0955033.
[44] Zhu, YB; Zhu, YQ. Growth property of silicon nanowires under OAG, *Int. J. Modern Phys. B,* 2005, 19, 683-685.
[45] Chueh, YL; Chou, LJ; Hsu, CM; Kung, SC. Synthesis and Characterization of Taper- and Rodlike Si Nanowires on SiXGe1-X Substrate. *J. Phys. Chem. B,* 2005, 109, 21831 -21835.
[46] Deng, CJ; Yu, P; Yau, MY; Ku, CS; Ng, DHL. Fabrication of single-crystal α-Al_2O_3 nanorods by displacement reactions. *J. Am. Ceram. Soc.,* 2003, 86, 1385-1388.
[47] Valcµrcel, V; PØrez, A; Cyrklaff, M; Guitiµn, F. Novel Ribbon-Shaped a-Al_2O_3 Fibers. *Adv. Mater.,* 1998, 10, 1370-1373.

[48] Peng, XS; Zhang, LD; Meng, GW; Wang, XF; Wang, YW; Wang, CZ; Wu, GS. Photoluminesence and infrared properties of α-Al2O3 nanowires and nanobelts. *J. Phys. Chem. B* 2002, 106, 11163-11167.

[49] Xu, XD; Wang, YH; Liu, ZF; Zhao, RG. A New Route to Large-Scale Synthesis of Silicon Nanowires in Ultrahigh Vacuum. *Adv. Funct. Mater.* 2007, 17, 1729–1734.

[50] Tang, YB; Cong, HT; Li, F; Cheng, HM. Synthesis and photoluminescent property of AlN nanobelt array. *Diamond Related Mater.,* 2007, 16, 537-541.

[51] Zhang, YF; Tang, YH; Wang, N; Lee, CS; Bello, I; Lee, ST. Germanium nanowires sheathed with an oxide layer. *Phys. Rev. B,* 2000, 61, 4518-4521.

[52] Shi, WS; Zheng, YF; Wang, N; Lee, CS; Lee, ST. A General Synthetic Route to III-V Compound Semiconductor Nanowires. *Adv. Mater.,* 2001, 13, 591-594.

[53] Hua, Q; Ma, XL; Xie, ZY; Wong, NB; Lee, CS; Lee, ST. Characterization of zinc oxide crystal whiskers grown by thermal evaporation, *Chem. Phys. Lett.,* 2001, 344, 97-100.

[54] Pan, ZW; Dai, ZR; Wang, ZL. Nanobelts of Semiconducting Oxides. *Science*, 2001, 291, 1947-1949.

[55] Ye, CH; Meng, GW; Jiang, Z; Wang, YH; Wang, GZ; Zhang, LD. Rational Growth of Bi2S3 Nanotubes from Quasi-two-dimensional Precursors. *J. Am. Chem. Soc.,* 2002, 124, 15180-15181.

[56] Wang, WZ; Zeng, BQ; Yang, J; Poudel, B; Huang, JY; Naughton, MJ; Ren, ZF. Aligned ultralong ZnO nanobelts and their enhanced field Eeission. *Adv. Mater.*, 2006, 18, 3275–3278.

[57] Wong, TC; Yu, CC; Wu, JJ. Low temperature formation of well-aligned nanocrystalline Si/SiOx composite nanowires. *Adv. Funct. Mater.,* 2005, 15, 1440-1444.

[58] He, JH; Yang, RS, Chueh, YL; Chou, LJ; Chen, LJ; Wang, ZL. Aligned AlN nanorods with multi-tipped surfaces-growth, field-emission, and cathodoluminescence properties. *Adv. Mater.*, 2006, 18, 650–654.

[59] Ouyang, L; Thrall, ES; Deshmukh, MM. Park, HK. Vapor-phase synthesis and characterization of e-FeSi nanowires. *Adv. Mater.,* 2006, 18, 1437–1440.

[60] Kim, C; Gu, WH; Briceno, M; Robertson, IM; Choi, H; Kim, K. Copper nanowires with a five-twinned structure grown by chemical vapor deposition. *Adv. Mater.,* 2008, 20, 1859–1863.

[61] Fang, XS; Ye, CH; Zhang, LD; Xie, T. Twinning-mediated growth of Al_2O_3 nanobelts and their enhanced dielectric responses. *Adv. Mater.,* 2005, 17, 1661-1665.

[62] Ksiazek, M; Sobczak, N; Mikulowski, B; Radziwill, W; Surowiak, I. Wetting and bonding strength in Al/Al_2O_3 system. *Mater.Sci. Eng. A,* 2002, 324, 162–167.

[63] Zardo, I; Yu, L; Conesa-Boj, S; Estrade, S; Alet, PJ; Roessler, J; Frimmer, M; Cabarrocas, PRI; Peiro, F; Arbiol, J; Morante, JR; Morral, AFI. Gallium assisted plasma enhanced chemical vapor deposition of silicon nanowires. *Nanotechnol.*, 2009, 20, Art. No 155602 (9pp).

[64] Hou, WC; Hong, FCN. Controlled surface diffusion in plasma-enhanced chemical vapor deposition of GaN nanowires. *Nanotechnol.* , 2009, 20, Art. No 055606 (6pp).

[65] Borras; A; Aguirre, M; Groening, O; Lopez-Cartes, C; Groening P. Synthesis of supported single-crystalline organic nanowires by physical vapor deposition. *Chem. Mater.* 2008, *20,* 7371–7373.

[66] Pang, JB; Qiu, KY; Wei, Y; Recent progress in research on mesoporous materials II: Application. *J. Inorg. Mater.,* 2002, 17, 665-671.

[67] Ikegame, M; Tajima, K; Aida, T. Template synthesis of polypyrrole nanofibers insulated within one-dimensional silicate channels: hexagonal versus lamellar for recombination of polarons into bipolarons. *Angew. Chem. Int. Ed.,* 2003, 42, 2154-2157.

[68] Rice, RL; Arnold, DC; Shaw, MT; Iacopina, D; Quinn, AJ; Amenitsch, H; Holmes, JD; Morris, MA. Ordered mesoporous silicate structures as potential templates for nanowire growth. *Adv. Funct. Mater.,* 2007, 17, 133–141.

[69] Fukuoka,A; Sakamoto,Y. Guan, SY; Inagaki, S; Sugimoto, N; Fukushima, Y; Hirahara,K; Iijima,S; Ichikawa, M. Novel templating synthesis of necklace-shaped mono- and bimetallic nanowires in hybrid organic-inorganic mesoporous material. *J. Am. Chem. Soc.,* 2001, 123, 3373 -3374.

[70] Coleman, NRB; O'Sullivan, N; Ryan, KM; Crowley, TA; Morris, MA; Spalding, TR; Steytler, DC; Holmes, JD. Synthesis and characterization of dimensionally ordered semiconductor nanowires within mesoporous silica. *J. Am. Chem. Soc.,* 2001, 123, 7010-7016

[71] Gao, F. Lu, QY; Liu, XY; Yan, YS; Zhao, DY. Controlled synthesis of semiconductor PbS nanocrystals and nanowires Inside Mesoporous Silica SBA-15 Phase. *Nano Lett.,* 2001, 1, 743 -748.

[72] Liu, XY; Tian, BZ; Yu, CZ, Tu, B; Liu, Z; Terasaki,O; Zhao, DY. Ordered nanowire arrays of metal sulfides templated by mesoporous silica SBA-15 via a simple impregnation reaction. *Chem. Letter.,* 2003, 824-825.

[73] Gao, F; Lu, QY; Zhao, DY. Synthesis of crystalline mesoporous CdS semiconductor nanoarrays through a mesoporous SBA-15 silica template technique. *Adv. Mater.,* 15, 739-742

[74] Han, YJ; Kim, JM; Stucky, GD. Preparation of noble metal nanowires using hexagonal mesoporous silica SBA-15. *Chem. Mater.,* 2000, 12, 2068-2069.

[75] Gu, JL; Shi, JL; Chen, HR; Xiong, LM; Shen, WH; Ruan, ML. Periodic pulse electrodeposition to synthesize ultra-high density CdS nanowire arrays templated by SBA-15 mesoporous films. *Chem. Lett.,* 2004, 828-829.

[76] Yang, CT; Huang, MH. Formation of arrays of gallium nitride nanorods within mesoporous silica SBA-15. *J. Phys. Chem. B,* 2005, 109, 17842-17847.

[77] Li, N; Li, XT; Geng, WC; Zhao, L; Zhu, GS; Wang, RW; Qiu, SL. Template synthesis of bron nitride nanotubes in mesoporous silica SBA-15. *Mater. Lett.,* 2005, 59, 925-928.

[78] Shi, KY; Chi, YJ; Yu, HT; Xin, BF; Fu, HG. Controlled growth of mesostructured crystalline iron oxide nanowires and Fe-filled carbon nanotube arrays templated by mesoporous silica SBA-16 film. *J. Phys. Chem. B,* 2005, 109, 2546-2551.

[79] Zhang, WX; Cui, JC; Tao, CA; Lin, CX; Wu, YG; Li, GT. Confined self-sssembly approach to produce ultrathin carbon nanofibers. *Langmuir,* 2009, 25, 8235–8239.

[80] Zhang, F; Wong, SS. Controlled synthesis of semiconducting metal sulfide nanowires. *Chem. Mater.,* 2009, 21, 4541–4554.

[81] Shin, HJ; Jeong, DK; Lee, J; Sung, MM; Kim, J. Formation of TiO_2 and ZrO_2 nanotubes using atomic layer deposition with ultraprecise control of wall thickness. *Adv. Mater.,* 2004, 16, 1197-1199.

[82] Mao, YB; Zhang, F; Wong, SS. Ambient template-directed synthesis of single-crystalline alkaline-earth metal fluoride nanowires. *Adv. Mater.,* 2006, 18, 1895–1899.

[83] Jessensky, O; Muller, F; Gosele, U. Self-organized formation of hexagonal pore structures in anodic alumina. *J. Electrochem. Soc.,* 1998, 145, 3735-3740.

[84] Li, AP; Müller, F; Birner, A; Nielsch, K; Gösele, U. Hexagonal pore arrays with a 50-420 nm interpore distance formed by self-organization in anodic alumina. *J. Appl. Phys.,* 1998, 84, 6023-6036.
[85] Qu, LT; Shi, GQ; Wu, XF; Fan, B. Facile route to silver nanotubes. *Adv. Mater.,* 2004, 16, 1200-1203.
[86] Zhang, SH; Xie, ZX; Jiang, ZY; Xu, X; Xiang, J; Huang, RB; Zheng, LS. Synthesis of silver nanotubes by electroless deposition in porous anodic aluminium oxide templates. *Chem. Commun.,* 2004, 1106.
[87] Mu, YY; Liang, HP; Hu, JS; Wan, LJ. A simple route to platinum and Pt-based composite nanotubes. *J. Nanosci. Nanotechnol.,* 2005, 5, 1929-1932.
[88] Kijima, T; Yoshimura, T; Uota, M; Ikeda, T; Fujikawa, D; Mouri, S; Uoyama, S. Noble-metal nanotubes (Pt, Pd, Ag) from lyotropic mixed-surfactant liquid-crystal templates. *Angew. Chem., Int. Ed.,* 2004, 43, 228-232.
[89] Steinhart, M; Jia, ZH; Schaper, AK; Wehrspohn, RB; Gosele, U; Wendorff, JH. Palladium nanotubes with tailored wall morphologies. *Adv. Mater.,* 2003, 15, 706-709.
[90] Ma, H; Tao, ZL; Gao, F; Chen, J. Template synthesis of transition-metal Ru and Pd nanotubes. *Chin. J. Inorg. Chem.,* 2004, 20, 1187-1190.
[91] lahav, M; Sehayek T; Vaskevich, A; Rubinstein, I. Nanoparticle nanotubes, Angew. *Chem. Int. Ed.,* 2003, 42, 5575-5579.
[92] Lee, M; Hong, S; Kim, D. Chemical-free synthesis of electrically connected gold nanotubes/nanoparticles from solution-infiltrated anodized aluminum oxide template. *Appl. Phys. Lett.,* 2006, 89, Art. No. 043120.
[93] Hanaoka, TA; Heilmann, A; KrPll, M; Kormann, HP; Sawitowski, T; Schmid, G; Jutzi, P; Klipp, A; Kreibig, U; Neuendorf, R. *Appl. Organomet. Chem.*, 1998, 12, 367 – 373.
[94] Lahav, M; Sehayek, T; Vaskevich, A; Rubinstein, I. Nanoparticle Nanotubes. *Angew.Chem. Int. Ed.,* 2003, 42, 5576 – 5579.
[95] Johansson, A; Lu, J; Carlsson, JO; Boman, M. Deposition of palladium nanoparticles on the pore walls of anodic alumina using sequential electroless deposition. *J. Appl. Phys.*, 2004, 96, 5189 – 5194.
[96] Barreca, D; Gasparotto, A; Tondello, E. Gold nanotubes by template-directed synthesis. *J. Nanosci. Nanotechnol.,* 2005, 5, 994-998.
[97] Lee, W; Scholz, R; Nielsch, K; Gosele, U. A template-based electrochemical method for the synthesis of multisegmented metallic nanotubes. *Angew. Chem. Int. Ed.,* 2005, 44, 6050 –6054.

[98] Mu, C; Yu, YX; Wang, RM; Wu, K; Xu, DS; Guo, GL. Uuniform metal nanotube arrays by multistep template replication and electrodeposition. *Adv. Mater.,* 2004, 16, 1550-1553.

[99] Sehayek, T; Lahav, M; Popovitz-Biro, R; Vaskevich, A; Rubinstein, I. Template synthesis of nanotubes by room-temperature coalescence of metal nanoparticles. *Chem. Mater.*, 2005, 17, 3743-3748.

[100] Tan, H; Ye, NY; Fan, WY. Alumina-template synthesis of fluorescent RuO_2 nanotubes derived from $Ru_3(CO)_{12}$ clusters. *Adv. Mater.,* 2006, 18, 619-623.

[101] Imai, H; Takei, Y; Shimizu, K; Matsuda M; Hirashima, HJ. Direct preparation of anatase TiO_2 nanotubes in porous alumina membranes. *Mater. Chem.,* 1999, 9, 2971 – 2972.

[102] Che, G; Lakshimi, BB; Martin, CR; Fisher, ER; Ruoff, RS. Chemical Vapor deposition based synthesis of carbon nanotubes and nanofibers using a template method. *Chem. Mater.,* 1998, 10, 260-267.

[103] Zhi, LJ; Wu, JS; Li, JX; Kolb, U; Mullen, K. Carbonization of disclike molecules in porous alumina membranes: Toward carbon nanotubes with controlled graphene-layer orientation. *Angew. Chem. Int. Ed.,* 2005, 44, 2120-2123.

[104] Shao, XF; Wu, XL; Huang, GS; Qiu, T; Jiang, M; Hong, JM. Alumina nanotubes and nanowires from Al-based porous alumina membranes. *Appl. Phys. A-Mater. Sci. Processing.,* 2005, 81, 621-625.

[105] Xiao, ZL; Han, CY; Welp, U; Wang, HH; Kwok, WK; Willing, GA; Hiller, JM; Cook, RE; Miller, DJ; Crabtree, GW. Fabrication of Alumina Nanotubes and Nanowires by Etching Porous Alumina Membranes. *Nano Lett.,* 2002, 2, 1293 –1297.

[106] Xu, QL; Meng,GW; Wu, XB; Wei, Q; Kong, MG; Zhu, XG; Chu, ZQ. A generic approach to desired metallic nanowires inside native porous alumina template via redox reaction. *Chem. Mater.,* 2009, 21, 2397–2402.

[107] Wu, B; Heidelberg, A; Boland, JJ. Mechanical properties of iltrahigh-strength gold nanowires, *Nature Mater.*, 2005, 4, 525-529.

[108] Yi, JB; Pan, H; Lin, JY; Ding, J; Feng, YP; Thongmee, S; Liu, T; Gong, H; Wang, L. Ferromagnetism in ZnO nanowires derived from electro-deposition on AAO template and subsequent oxidation. *Adv. Mater.,* 2008, 20, 1170–1174

[109] Cao, HQ; Qiu, XQ; Luo, B; Liang, Y; Zhang, YH; Tan, RQ; Zhao, MJ; Zhu, QM. Synthesis and room temperature ultraviolet

photoluminescence properties of zirconia nanowires. *Adv. Funct. Mater.,* 2004, 14, 243-246.

[110] Li, CY; Guo, YG; Li, BS; Wang, CR; Wan, LJ; Bai, CL. Template synthesis of SC@C-82(I) nanowires and nanotubes at room. *Adv. Mater.,* 2005, 17, 71-73.

[111] Bocchetta, P; Santamaria, M; Di, QF. Template electrosynthesis of $La(OH)_3$ and $Nd(OH)_3$ nanowires using porous anodic alumina membranes. *Electrochem. Comm.,* 2007, 9, 683-688.

[112] Matsui, K; Kyotani, T; Tomita, A. Hydrothermal synthesis of sibgle-crystal $Ni(OH)_2$ nanorods in a carbon-coated anodic alumina film. *Adv. Mater.,* 2002, 14, 1216-1219.

[113] Jha, H; Kikuchi, T; Sakairi, M; Takahashi, H. Synthesis of aluminum oxy-hydroxide nanofibers from porous anodic alumina. *Nanotechnol.* , 2008, 19, Art. No. 395603 (6pp).

[114] Feng, L; Li, SH; Zhai, J; Song, YL; Jiang, L; Zhu, DB. Template based synthesis of aligned polyacrylonitrile nanofibers using a novel extrusion method. *Synthetic Metal,* 2003, 135-136, 817-818.

[115] Carretero-Genevrier, A; Mestres, N; Puig, T; Hassini, A; OrÓ, J; Pomar, A; Sandiumenge, F; Obradors, X; Ferain, E. Single-crystalline La0.7Sr0.3MnO3 nanowires by polymer-template-directed chemical solution synthesis. *Adv. Mater.,* 2008, 20, 3672–3677.

[116] A special issue on carbon nanotubes. *Acc Chem Res.,* 2002, 35, 997-1113.

[117] Ajayan, P M. Nanotubes from Carbon. *Chem. Rev.,* 1999, 99, 1787-1800.

[118] Tasis, D; Tagmatarchis, N; Bianco, A; Prato, M. Chemistry of Carbon Nanotubes. *Chem. Rev.,* 2006, 106, 1105 –1136.

[119] Barros, EB; Jorio, A; Samsonidze, GG; Capaz, RB; Souza, AG; Mendes, J; Dresselhaus, G; Dresselhaus, MS. Review on the symmetry-related properties of carbon nanotubes. *Phys. Report-review section Phys. Lett.,* 2006, 431, 261-302.

[120] Gibson, RF; Ayorinde, EO; Wen, YF. Vibrations of carbon nanotubes and their composites: A review. *Composites Sci Tech.,* 2007, 67, 1-28.

[121] Iijima, S. Helical microtubules of graphitic carbon. *Nature,* 1991, 354, 56-58.

[122] Han, WQ; Fan, SS; Li, QQ; Hu, YD. Synthesis of gallium nitride nanorods Through a carbon nanotube–confined reaction. *Science,* 1997, 277, 1287-1289.

[123] Dai, HJ; Wong, EW; Lu, YZ; Fan, SS; Lieber, CM. Synthesis and characterization of carbide nanorods. *Nature,* 1995, 375, 769-772.

[124] Tang, CC; Fan, SS; de la Chapelle, ML; Dang, HY; Li, P. Synthesis of gallium phosphide nanorods. Adv. Mater. 2000, 12, 1346-1348.

[125] Ye, XR; Lin, YH; Wang, CM; Wai, CM. Supercritical fluid fabrication of metal nanowires and nanorods Templated by multiwalled carbon nanotubes. *Adv. Mater.,* 2003, 15, 316-319.

[126] Sun, ZY; Yuan, HQ; Liu, ZM; Han, BX; Zhang, XR. A highly efficient chemucal sensor materials for H_2S: α-Fe_2O_3 nanotubes fabricated using carbon nanotube templates. *Adv. Mater.,* 2005, 17, 2993-2997.

[127] O'Connell, MJ; Boul, P; Ericson, LM; Huffman, C; Wang, Y; Haroz, E; Kuper, C; Tour, J; Ausman, KD; Smalley, RE. *Chem. Phys. Lett.,* 2001, 342, 265.

[128] Correa-Duarte, MA; Sobal, N; Liz-Marzan, LM; Giersig, M. Linear assemblies of silica-coated gold nanoparticles using carbon nanotubes as templates. *Adv. Mater.,* 2004, 16, 2179-2183.

[129] Lazzara, TD; Bourret, GR; Lennox, RB; van de Ven TGM. Polymer templated synthesis of AgCN and Ag nanowires. *Chem. Mater.* **2009,** *21,* 2020–2026

[130] Nam, KT; Kim, DW; Yoo, PJ; Chiang, CY; Meethong, N; Hammond, PT; Chiang, YM; Belcher, AM. Virus-enabled synthesis and assembly of nanowires for lithium ion battery electrodes. *Science,* 2006, 312, 885-888.

[131] Hopkins, DS; Pekker, D; Goldbart, PM; Bezryadin, A. Quantum interference device made by DNA templating of superconducting nanowires. *Science,* 2005, 308, 1762-1765.

[132] Dong, LQ; Hollis, T; Connolly, BA; Wright, NG; Horrocks, BR; Houlton. *A.* DNA templated semiconductor nano- particle chains and wires. *Adv. Mater.,* 2007, 19, 1748-1751.

[133] Atanasova, P; Weitz, RT; Gerstel, P; Srot, V; Kopold, P; van Aken, PA; Burghard, M; Bill, J; DNA-templated synthesis of ZnO thin layers and nanowires. *Nanotechnol.*, 2009, 20, Art. No. 365302 (6pp).

[134] Yuan, JY; Xu, YY; Walther, A; Bolisetty, S; Schumacher, M; Schmalz, H; Ballauff, M; Műller, AHE. Water-soluble organo-silica hybrid nanowires. *Nature Mater*. 2008, 7, 718-722.

[135] Wang, H; Patil, AJ; Liu, K; Petrov, S; Mann, S; Winnik, MA; Manners, I. Fabrication of continuous and segmented polymer/metal oxide nanowires using cylindrical micelles and block comicelles as templates. *Adv. Mater.,* 2009, 21, 1805–1808

[136] Bognitzki, M.; Hou, H. Q.; Ishaque, M.; Frese, T.; Hellwig, M.; Schwarte, C.; Schaper, A.; Wendorff, J. H.; Greiner, A. Polymer, metal, and hybrid nano- and mesotubes by coating degradable polymer template fibers (TUFT process). *Adv. Mater.,* 2000, 12, 637-640.

[137] Caruso, RA; Schattka, JH.; Greiner, A. Titanium dioxide tubes from sol-gel coating of electrospun polymer fibers. *Adv. Mater.,* 2001, 13, 1577-1579.

[138] Wang, Y; Qin, Y; Berger,A; Yau, E; He,CC; Zhang, LB; Gösele, U; Knez, M; Steinhart, M. Nanoscopic morphologies in block copolymer nanorods as templates for atomic-layer deposition of semiconductors. *Adv. Mater.,* 2009, 21, 2763–2766.

[139] Santala, E; Kemell, M; Leskela, M; Ritala, M; The preparation of reusable magnetic and photocatalytic composite nanofibers by electrospinning and atomic layer deposition. *Nanotechnol.*, 2009. Art. No 035602 (6pp).

[140] Park, K.; Lee, JS; Sung, MY; Kim, S. Structural and optical properties of ZnO nanowires synthesized from ball-milled ZnO powders. *Jpn. J. Appl. Phys.*, 2002, 41, 7317-7321.

[141] Hwang, J; Min, B; Lee, JS; Keem, K; Cho, K; Sung, MY; Lee, MS; Kim, S. Al_2O_3 nanotubes fabrication by wet etching of ZnO/Al_2O_3 core/shell nanofibers. *Adv. Mater.,* 2004, 16, 422-425.

[142] Chiu, JJ; Kei, CC; Wang, WS; Perng, TP. Organic semiconductor nanowires for field emission. *Adv. Mater.,* 2003, 15, 1361-1364.

[143] Wang, CC; Kei, CC; Yu, YW; Perng, TP. Organic Nanowire-Templated Fabrication of Alumina Nanotubes by Atomic Layer Deposition. *Nano Lett.,* 2007, 7, 1566-1569.

[144] Ogihara, H; Sadakane, M; Nodasaka, Y; Ueda, W. Shape-Controlled Synthesis of ZrO_2, Al_2O_3, and SiO_2 Nanotubes Using Carbon Nanofibers as Templates. *Chem. Mater.,* 2006, 18, 4981-4983.

[145] Ogihara, H; Masahiro, S; Nodasaka, Y; Ueda, W. Synthesis,characterization and formation process of transition metal oxide nanotubes using carbon nanofibers as templates. *J. Solid State Chem.,* 2009, 182, 1587–1592.

[146] Xie, GW; Wang, ZB; Li, GC; Shi, YL; Cui, ZL; Zhang, ZK. Templated synthesis of metal nanotubes via electroless deposition. *Mate. Lett.,* 2007, 61, 2641–2643.

[147] Niu, H; Gao, M. Diameter-tunable CdTe nanotubes tmplated by 1D nanowires of Cadminum Thiolate polymer. *Angew. Chem. Int. Ed.*, 2006, 45, 6462-6466.

[148] Cui, XJ; Yu, SH; Li, LL: Li, K; Yu, B. Fabrication of Ag_2SiO_3/SiO_2 composite nanotubes using one-step sacrificial templating solution approach. *Adv. Mater.,* 2007, 16, 1109-1112.

[149] Fan, X; Meng, XM; Zhang, XH; Lee, CS; Lee, ST. Template fabrication of SiO_2 nanotubes. *Appl. Phys. Lett.,* 2007, 90, 103114-103114.

[150] Choi, SH; Ankonina, G; Youn, DY; Oh, SG; Hong, JM; Rothschild, A; Kim. ID. Hollow ZnO nanofibers fabricated using electrospun polymer templates and their electronic transport properties. *ACS NANO*, 2009, 3, 2623–2631.

[151] Li, YB; Bando, Y; Golberg, D. Single-crystalline α-Al_2O_3 nanotubes converted from Al_4O_4C nanowires. *Adv. Mater.,* 2005, 17, 1401-1405.

[152] *Gautam*, UK; *Bando*, Y; *Zhan*, JH; *Costa*, PMFJ; *Fang*, XS; *Golberg, D*. Ga-Doped ZnS Nanowires as Precursors for $ZnO/ZnGa_2O_4$ Nanotubes. *Adv. Mater.,* 2008, 20, 810–814.

[153] Elias, J; Tena-Zaera, R; Wang, GY; Lévy-Clément, C. Conversion of ZnO nanowires into nanotubes with tailored dimensions. *Chem. Mater.,* 2008, *20,* 6633–6637.

[154] [1]Cao, MH; Wang, YH; Guo, CX; Qi, YJ; Hu, CW. Preparation of ultrahigh-aspect-ratio hydroxyapatite nanofibers in reverse micelles under hydrothermal conditions. *Langmuir*, 2004, 20, 4784-4786.

[155] Wang, L; Yuan, ZZ; Chen, QH; Sun, F; Zhu, LC. Synthesis of prismatic single crystal gamma-MnOOH. *J. Inorg. Mater.,* 2007, 22, 667-670.

[156] Fang, YP; Xu, AW; You, LP; Song, RQ; Yu, JC; Zhang, HX; Li, Q; Liu, HQ. Hydrothermal synthesis of rare earth (Tb, Y) hydroxide and oxide nanotubes. *Adv. Mater.,* 2003, 13, 955-960.

[157] Teng, F; Han, W; Liang, SH; Gaugeu, BG; Zong, RL; Zhu, YF. Catalytic behavior of hydrothermally synthesized $La_{0.5}Sr_{0.5}MnO_3$ single-crystal cubes in the oxidation of CO and CH_4. *J. Catal.,* 2007, 250, 1-11.

[158] Arnold, DC; Kazakova, O; Audoit, G; Tobin, JM; Kulkarni, JS; Nikitenko,S; Morris, MA; Holmes, JD. The synthesis and characterisation of ferromagnetic CaMn2O4 nanowires. *Chem. Phys. Chem.,* 2007, 8, 1694-1700.

[159] Caswell, KK; Bender, CM; Murphy, CJ. Seedless, surfactantless wet chemical synthesis of silver nanowires. *Nano Lett.,* 2003, 3, 667-669.

[160] Lu, QY; Gao, F; Zhao, DY. One-step synthesis and assembly of copper sulfide nanoparticles to nanowires, nanotubes, and nanovesicles by a simple organic amine-assisted hydrothermal process. *Nano Lett.,* 2002, 2, 725-728.

[161] Zhang, H; Yang, DR; Li, SZ; Ji, YJ; Ma, XY; Que, DL. Hydrothermal synthesis of flower-like Bi2S3 with nanorods in the diameter region of 30 nm. *Nanotechnol.,* 2004, 15, 1122-1125.

[162] Therese, HA; Li, JX; Kolb, U; Tremel, W. Facile large scale synthesis of WS_2 nanotubes from WO3 nanorods prepared by a hydrothermal route. *Solid State Sci.,* 2005, 7, 67-72.

[163] Zhang, H; Ji, YJ; Ma, XY; Xu, J; Yang, DR. Long Bi_2S_3 nanowires prepared by a simple hydrothermal method. *Nanotechnol.,* 2003, 14, 974-977.

[164] Gu, ZJ; Ma, Y; Zhai, TY; Gao, BF; Yang, WS. A simple hydrothermal method for the large-scale synthesis of single-crystal potassium tungsten bronze nanowires. *Chem-A Euro. J.,* 2006, 12, 7717-7723.

[165] Reis, KP; Ramanan, A; Whittingham, MS. Hydrothermal Synthesis of Sodium Tungstates. *Chem. Mater.,* 1990, 2, 219-221.

[166] Wang, HL; Ma, XD; Qian, XF; Yin, J; Zhu, ZK. Selective synthesis of $CdWO_4$ short nanorods and nanofibers and their self-assembly. *J. Solid State Chem.,* 2004, 177, 4588-4596.

[167] Song, XC; Zheng, YF; Wang, Y; Cao, GS; Yin, HY. Na_3PO_4 assisted hydrothermal synthesis of WO_3 nanorods. *J. Inorg. Mater.*, 2006, 21, 1472-1476.

[168] Cui, XJ; Yu, SH; Li, LL; Biao, L; Li, HB; Mo, MS; Liu, XM. Selective synthesis and characterization of single-crystal silver molybdate/tungstate nanowires by a hydrothermal process. *Chem-A Euro. J.,* 2004, 10, 218-223.

[169] Panda, AB; Acharya, S; Efrima, S. Ultranarrow ZnSe nanorods and nanowires: structure, spectroscopy, and one-dimensional properties. *Adv. Mater.,* 2005, 17, 2471–2474.

[170] Purkayastha, A; Lupo, F; Kim, S; Borca-Tasciuc, T; Ramanath, G. Low-temperature, template-free synthesis of single-crystal bismuth telluride nanorods. *Adv. Mater.*, 2006, 18, 496–500.

[171] Chen, YW; Tang, YH; Pei LZ; Guo, C. Self-assembled silicon nanotubes grown from silicon monoxide. *Adv. Mater.,* 2005, 17, 564-567

[172] Lin, LW; Tang, YH; Li, XX; Pei, LZ; Guo, C. Water-assisited synthesis of silicon oxide nanowires under supercritical hydrothermal conditions. *J. Appl. Phys.,* 2007, 101, 014314-1-014314-7.

[173] Gao, QS; Chen, P; Zhang, YH; Tang, Y. Synthesis and characterization of organic–inorganic hybrid GeOx/ethylenediamine nanowires. *Adv. Mater.*, 2008, 20, 1837–1842.

[174] Li, YY; Liu, JP; Ha, ZJ. Fabrication of boehmite AlOOH nanofibers by a simple hydrothermal process. *Mater. Lett.,* 2006, 60, 3596-3590.

[175] Hou, HW; Xie, Y; Yang, Q; Guo, QX; Tan, CR. Preparation and characterization of gamma-AlOOH nanotubes and nanorods. *Nanotechnol.,* 2005, 16, 741-745.

[176] Chen, XY; Lee, SW. pH-dependent formation of boehmite (gamma-AlOOH) nanorods and nanoflakes. *Chem. Phys. Lett.,* 2007, 438, 279-284.

[177] Dou, QS; Zhang, H; Wu, JB; Yang, DR. Synthesis and characterization of Fe_2O_3 and FeOOH nanostructures prepared by ethylene glycol assisted hydrothermal process. *J. Inorg. Mater.*, 2007, 22, 213-218.

[178] Liu, XH; Qiu, GZ; Yan, AG; Wang, Z; Li, XG.Hydrothermal synthesis and characterization of alpha-FeOOH and alpha-Fe_2O_3 uniform nanocrystallines. *J. Alloys Compounds,* 2007, 433, 216-220.

[179] Matsui, K; Kyotani, T; Tomita, A. Hydrothermal synthesis of single-crystal Ni(OH)(2) nanorods in a carbon-coated anodic alumina film. *Adv. Mater.,* 2002, 14, 1216-1219.

[180] Li, XL; Liu, JF; Li, YD. Low-temperature conversion synthesis of $M(OH)_2$ (M = Ni, Co, Fe) nanoflakes and nanorods. Mater. *Chem. Phys.*, 2003, 80, 222-227.

[181] Jiao, QZ; Tian, ZL; Zhao, Y. Preparation of nickel hydroxide nanorods/nanotubes and microscopic nanorings under hydrothermal conditions. *J. Nanoparticle Res.,* 2007, 9, 519-522.

[182] Lu, CH; Wang, HC. Formation and microstructural variation of cerium carbonate hydroxide prepared by the hydrothermal process. *Mater. Sci. Eng. B-Solid State Mater. Adv. Tech.,* 2002, 90, 138-141.

[183] Tang, CC; Bando, Y; Liu, BD; Golberg, D. Cerium oxide nanotubes prepared from cerium hydroxide nanotubes. *Adv. Mater.,* 2005, 17, 3005- 3009.

[184] Yin, YD; Hong, GY. Synthesis and characterization of $La(OH)_3$ nanorods by hydrothermal microemulsion method. *Chin. Chem. Lett.,* 2005, 16, 1659-1662.

[185] Deng, Y; Wu, JB; Liu, J; Wei, GD; Nan, CW. Hydrothermal growth and characterization of $La(OH)_3$ nanorods and nanocables with $Ni(OH)_2$ coating. *J Phys. Chem. Solid,* 2003, 64, 607-610.

[186] Dong, XY; Zhang, XT; Liu, B; Wang, HZ; Li, YC; Huang, YB; Du, ZL. Controlled synthesis of manganese oxohydroxide (MnOOH) and Mn3O4 nanorods using novel reverse micelles. *J. Nanosci. Nanotechnol.*, 2006, 6, 818-822.

[187] Zhang, YC; Qiao, T; Hu, XY; Zhou, WD. Simple hydrothermal preparation of gamma-MnOOH nanowires and their low-temperature thermal conversion to beta-MnO_2 nanowires. *J. Cryst. Grow.,* 2005, 280, 652-657.

[188] Hu, CG; Liu, H; Dong, WT; Zhang, YY; Bao, G; Lao, CS; Wang, ZL. $La(OH)_3$ and La_2O_3 nanobelts—synthesis and physical properties. *Adv. Mater.,* 2007, 19, 470–474.

[189] Tian, ZB; Feng, Q; Sumida, N; makita, Y; Ooi, K. Synthesis of manganese oxide nanofibers by selfassembling hydrothermal process. *Chem. Lett.,* 2004, 33, 952-953.

[190] Vayssieres, L; Sathe, C; Butorin, SM; Shuh, D.K; Nordgren J; Guo, JH. One-dimensional quantum-confinement effect in α-Fe_2O_3 ultrafine nanorods arrays. *Adv. Mater.,* 2005, 17, 2320-2323.

[191] Zhu, HL; Yang, DR; Zhang, H. A simple and novel low-temperature hydrothermal synthesis of ZnO nanorods. *Inorg. Mater.,* 2006, 42, 1210-1214.

[192] Vayssieres, L. Growth of arrayed nanorods and nanowires of ZnO from aqueous solution. *Adv. Mater.,* 2003, 15, 464-466.

[193] Fan, LB; Song, HW; Li, T; Yu, LX; Liu, ZX; Pan, GH; Lei, YQ. Bai, X; Wang, T; Zheng, ZH; Kong, XG. Hydrothermal synthesis and photoluminescent properties of ZnO nanorods. *J. Luminescence,* 2007, 122, 819-821.

[194] Tam, KH; Cheung, CK; Leung, YH; Djurisic, AB; Ling, CC; Beling, CD; Fung, S; Kwok, WM; Chan, WK; Phillips, DL; Ding, L; Ge, WK. Defects in ZnO nanorods prepared by a hydrothermal method. *J. Phys. Chem. B,* 2006, 110, 20865-20871.

[195] Xu, CX; Wei, A; Sun, XW; Dong, ZL. Aligned ZnO nanorods synthesized by a simple hydrothermal reaction. *J. Phys. D- Appl. Phys*., 2006, 39, 1690-1693.

[196] Le, HQ; Chua, SJ; Loh, KP; Fitzgerald, EA; Koh, YW. Synthesis and optical properties of well aligned ZnO nanorods on GaN by hydrothermal synthesis. *Nanotechnol.,* 2006, 17, 483-488.

[197] Kim, JH; Andeen, D; Lange, FF. Hydrothermal growth of periodic, single-crystal ZnO microrods and microtunnels. *Adv. Mater.,* 2006, 18, 2453–2457

[198] Lu, CH; Qi, LM; Yang, JH; Tang, L; Zhang, DY; Ma, JM. Hydrothermal growth of large-scale micropatterned arrays of ultralong ZnO nanowires and nanobelts on zinc substrate. *Chem. Commun.*, 2006, 3551-3553.

[199] Liu, B; Zeng, HC. Hydrothermal synthesis of ZnO nanorods in the diameter regime of 50 nm. *J. Am. Chem. Soc.,* 2003, 125, 4430-4431.

[200] Wang, X; Li, YD. Selected-control hydrothermal synthesis of γ- and α-MnO_2 single crystal nanowires. *J. Am. Chem. Soc.*, 2002, 124, 2880-2881.

[201] Zhang, YX; Li, GH; Jin, YX; Zhang, Y; Zhang J; Zhang, LD. Hydrothermal synthesis and photoluminescence of TiO_2 nanowires, Chem. Phys. Lett. 2002, 365, 300-304.

[202] Nian, JN; Teng, HS. Hydrothermal synthesis of single-crystalline anatase TiO_2 nanorods with nanotubes as the precursor. *J. Phys. Chem. B*, 2006, 110, 4193-4198.

[203] Ma, GB; Zhao, XN; Zhu, JM. Microwave hydrothermal synthesis of rutile TiO_2 nanorods. *Int. J. Morden Phys.,* 2005, 19, 2763-2768.

[204] Yoshida, R; Suzuki, Y; Yoshikawa, S. Syntheses of TiO_2(B) nanowires and TiO_2 anatase nanowires by hydrothermal and post-heat treatments. *J. Solid State Chem.,* 2005, 178, 2179-2185.

[205] Lou, XW; Zeng, HC. Hydrothermal synthesis of alpha-MoO_3 nanorods via acidification of ammonium heptamolybdate tetrahydrate. *Chem. Mater.,* 2002,14, 4781-4789.

[206] Lou, XW; Zeng, HC. An inorganic route for controlled synthesis of $W_{18}O_{49}$ nanorods and nanofibers in solution. *Inorg. Chem.,* 2003, 42, 6169-6171.

[207] Zhu, DL; Zhu, H; Zhang, YH. Hydrothermal synthesis of $La_{0.5}Ba_{0.5}MnO_3$ nanowires. *Appl. Phys. Lett.,* 2002, 80, 1634-1636.

[208] Zhu, HY; Gao, XP; Song, DY; Bai, YQ; Ringer, SP; Gao, Z; Xi, YX, Martens, W; Riches, JD; Frost, RL. Growth of Boehmite Nanofibers by Assembling Nanoparticles with Surfactant Micelles. *J. Phys. Chem. B,* 2004, 108, 4245-4247.

[209] Zhao, YY; Martens, WN; Bostrom, TE; Zhu, HY; Frost, RL. Synthesis, charactreization, and surface properties of iron-doped boehmite nanofibers. *Langmuir*, 2007, 23, 2110-2116.

[210] Zhu, HY; Riches, JD; Barry, JC. Gamma-alumina nanofibers prepared from aluminum hydrate with poly(ethylene oxide) surfactant. *Chem. Mater.,* 2002, 14, 2086-2093.

[211] Kuang, DB; Fang, YP; Liu, HQ; Frommen, C; Fenske, D. Fabrication of boehmite AlOOH and gamma-Al_2O_3 nanotubes via a soft solution route. *J. Mater. Chem.,* 2003, 13, 660-662.

[212] Song, RQ; Xu, AW; Deng, B; Fang, YP. Novel multilamellar mesostructured molybdenum oxide nanofibers and nanobelts: synthesis and characterization. *J. Phys. Chem. B*, 2005, 109, 22758-22766.
[213] Chu, DW; Zeng, YP; Jiang, DL. Preparation of ZnO nanorods via surfactant assisted hydrothermal synthesis. *J. Inorg. Mater.*, 2006, 21, 571-575.
[214] Wang, JF; Tsung, CK; Hong, WB; Wu,YY; Tang, J; Stucky, GD. Synthesis of mesoporous silica nanofibers with controlled pore architectures. *Chem. Mater.,* 2004, 16, 5169-5181.
[215] Zhang, ZT; Rodinone, AJ; Ma, JX; Shen, J; Dia, S. Morphologically templated growth of aligned spinel $CoFe_2O_4$ nanorods. *Adv. Mater.*, 2005, 17, 1415-1419.
[216] Yu, SH; Colfen, H; Antonietti, M. The combination of colloid-controlled heterogeneous nucleation and polymer controlled crystallization: facile synthesis of separated, uniform high-aspect-ratio single-crystalline $BaCrO_4$ nanofibers. *Adv. Mater.,* 2003, 15, 133-136.
[217] Li, XK; Chang, J. A novel hydrothermal route to the synthesis of xonotile nanofibers and investigation on their bioactivity. *J. Mater. Sci.,* 2006, 41, 4944-4947.
[218] Purkayastha, A; Yan, QY; Raghuveer, MS; Gandhi, DD; Li, HF; Liu, ZW; Ramanujan, RV; Borca-Tasciuc, T; Ramanath, G. Surfactant-directed synthesis of branched bismuth telluride/sulfide core/shell nanorods. *Adv. Mater.,* 2008, 20, 2679–2683.
[219] Wang, ZH; Liu, JW; Chen, XY; Wan, JX; Qian, YT. A Simple hydrothermal route to large-scale synthesis of uniform silver nanowires. *Chem. Eur. J.,* 2005, 11, 160 – 163.
[220] Xu, J; Hu, J; Peng, CJ; Liu, HL; Hu, Y. A simple approach to the synthesis of silver nanowires by hydrothermal process in the presence of gemini surfactant. *J. Coll. Intface Sci.,* 2006, 298, 689-693.
[221] Shi, Y; Li, H; Chen, LQ; Huang, XJ. Obtaining ultra-long copper nanowires via a hydrothermal process. *Sci. Tech. Adv. Mater.,* 2005, 6, 761-765.
[222] Liu, ZP; Yang, Y; Liang, JB; Hu, ZK; Li, S; Peng, S; Qian, YT. Synthesis of copper nanowires via a complex-surfactant-assisted hydrothermal reduction process. *J. Phys. Chem. B,* 2003, 107, 12658-12661.
[223] Zhang, JL; Du, JM; Han, BX; Liu, ZM; Jiang, T; Zhang, ZF. Sonochemical formation of single-crystalline gold nanobelts. *Angew. Chem. Int Ed.* 2006, 45, 1116-1119.

[224] Gerung, H; Boyle, TJ; Tribby, LJ; Bunge, SD; Brinker, CJ; Han, SM. Solution Synthesis of Germanium Nanowires Using a Ge^{2+} Alkoxide Precursor. *J. Am. Chem Soc.,* 2006, 128, 5244-5250.

[225] Xie, BQ; Qian, YT; Zhang, SY; Fu, SQ; Yu, WC. A hydrothermal reduction route to single-crystalline hexagonal cobalt nanowires . Euro. *J. Inorg. Chem.,* 2006, (12), 2454-2459.

[226] Liu, YK; Wang, WZ; Zhang, YJ; Zheng, CL; Wang, GH. A simple route to hydroxyapatite nanofibers. *Mater. Lett.,* 2002, 56, 496-501.

[227] Cao, MH; Wang, YH; Guo, CX; Qi, YJ; Hu, CW. Preparation of ultrahigh-aspect-ratio hydroxyapatite nanofibers in reverse micelles under hydrothermal conditions. *Langmuir*, 2004, 20, 4784-4786.

[228] Wang, X; Zhuang, J; Peng, Q; Li, YD. Liquid–Solid–Solution synthesis of biomedical hydroxyapatite nanorods. *Adv. Mater.,* 2006, 18, 2031–2034.

[229] Wang, F; Li, MS; Lua, YP; Qi, YX; Liu, YX. Synthesis and microstructure of hydroxyapatite nanofibers synthesized at 37°C. *Mater. Chem. Phys.,* 2006, 95, 145–149.

[230] Samal, AK; Pradeep, T; Room-temperature chemical synthesis of silver telluride nanowires. *J. Phys. Chem. C,* 2009, *113,* 13539–13544.

[231] Cheng, B; Samulski, ET. Hydrothermal synthesis of one-dimensional ZnO nanostructures with different aspect ratios. *Chem. Commun.,* 2004, (8) 986-987.

[232] Ayudhya, SKN, Tonto, P; Mekasuwandumrong, O; Pavarajarn, V; Praserthdam, P. Solvothermal synthesis of ZnO with various aspect ratios using organic solvents. *Cryst. Grow. Design,* 200, 6, 2446-2450.

[233] Liu B; Zeng, HC. Room Temperature solution synthesis of monodispersed Single-crystalline ZnO nanorods and derived hierarchical nanostructures. *Langmuir,* 2004, 20, 4196-4204.

[234] Lin, CF; Lin, H. Wang, N; Zhang, X; Yang, J. Li, JB; Yang, XZ. Facile synthesis of long, straight and uniform copper nanowires via a solvothermal method. *Rare Metal Mater. Eng.,* 2006, 35, 644-645.

[235] Sui, RH; Thangadurai, V; Berlinguette, CP. Simple protocol for generating TiO_2 nanofibers in organic media. *Chem. Mater.* ,2008, 20, 7022–7030.

[236] Yan, QY; Chen, H; Zhou, WW; Hng, HH; Boey, FYC; Ma , J. A Simple chemical approach for PbTe nanowires with enhanced thermoelectric properties. , *Chem. Mater.*, 2008, 20, 6298–6300.

[237] Jiang, Y; Wu, Y; Zhang, SY; Xu, CY; Yu, WC; Xie, Y; Qian, YT. A catalytic-assembly solvothermal route to multiwall carbon nanotubes at a moderate temperature. *J. Am. Chem. Soc.,* 2000, 122, 12383-12384.

[238] Wang, WZ; Kunwar, S; Huang, JY; Wang, DZ; Ren, ZF. Low temperature solvothermal synthesis of multiwall carbon nanotubes. *Nanotechnol.*, 2005, 16, 21-23.

[239] Zhang, W; Ma, DK; Liu, JW; Kong, LF; Yu, WC; Qian YT. Solvothermal synthesis of carbon nanotubes by metal oxide and ethanol at mild temperature. *Carbon*, 2004, 42, 2341-2343.

[240] Luo, T; Chen, LY; Bao, KY; Yu WC; Qian . YT. Solvothermal preparation of amorphous carbon nanotubes and Fe/C coaxial nanocables from sulfur, ferrocene, and benzene. *Carbon*, 2006, 44, 2844-2848.

[241] Mu, TC; Huang, J; Liu, ZM; Li, ZH; Han, BX. Solvothermal synthesis of carbon nitrogen nanotubes and nanofibers. *J. Mater. Res.*, 2006, 21, 1658-1663.

[242] Huang, FL; Cao, CB; Zhu, HS. Catalytic self-assembly preparation and characterization of carbon nitride nanotubes by a solvothermal method. *Chin. Sci. Bull.*, 2005, 50, 626-629.

[243] Holmes, JD; Johnston, KP; Doty, RC; Korgel, BA. Control of thickness and orientation of solution-grown silicon nanowires. *Science*, 2000, 287, 1471-1473.

[244] Tang, KB; Qian, YT; Zeng, JH; Yang, XJ. Solvothermal route to semiconductor nanowires. *Adv. Mater.*, 2003, 15, 448-450.

[245] Zou, GF; Li, H; Zhang, YG; Xiong, K; Qian, YT. Solvothermal/hydrothermal route to semiconductor nanowires. *Nanotechnol.*, 2006, 17, S313-S320.

[246] Datta, A; Kar, S; Ghatak, J; Chaudhuri, S. Solvothermal synthesis of CdS nanorods: Role of basic experimental parameters. *J. Nanosci. Nanotechnol.,* 2007, 7, 677-688.

[247] Cao, BL; Jiang, Y; Wang, C; Wang, WH; Wang, LZ; Niu, M; Zhang, WJ; Li, YQ; Lee, ST. Synthesis and lasing properties of highly ordered CdS nanowire arrays, *Adv. Funct. Mater.,* 2007, 17, 1501–1506.

[248] Yao, WT; Yu, SH; Liu, SJ; Chen, JP; Liu, XM; Li, FQ. Architectural control syntheses of CdS and CdSe nanoflowers, branched nanowires, and nanotrees via a solvothermal approach in a mixed solution and their photocatalytic property. *J. Phys. Chem.,* 2006, 110, 11704-11710.

[249] Xu, XX; Wei, W; Qiu, XM; Yu, KH; Yu, RB; Si, SM; Xu, GQ; Huang, W; Peng, B. Synthesis of InAs nanowires via a low-temperature solvothermal route. *Nanotechnol.*, 2006, 17, 3416-3420.

[250] Xu, XX; Yu, KH; Wei, W; Peng, B; Huang, SH; Chen, ZH; Shen, XS. Raman scattering in InAs nanowires synthesized by a solvothermal route. *Appl. Phys. Lett.,* 2006, 89, Art. No. 253117.

[251] Liu, ZP; Peng, S; Xie, Q; Hu, ZK; Yang, Y; Zhang, SY; Qian, YT. Large-scale synthesis of ultraBi2S3 nanoribbons via a solvothermal process. *Adv. Mater.,* 2003, 936-939.

[252] Yang, XH; Wang, X; Zhang, ZD. Facile solvothermal synthesis of single-crystalline Bi2S3 nanorods on a large scale. Mater. *Chem. Phys.,* 2006, 95, 154-157.

[253] Sigman, MB; Korgel, BA. Solvothermal synthesis of Bi_2S_3 (bismuthinite) nanorods, nanowires and nanofabic. *Chem. Mater.,* 2005, 17, 1655-1660.

[254] Shi, HQ; Zhou, XD; Xun, Fu; Wang, DB; Hu, ZS. Preparation of CdS nanowires and Bi2S3 nanorods by extraction–solvothermal method. *Mater. Lett.,* 2006,60, 1793-1795.

[255] Liu, XZ; Cui, JH; Zhang, LP; Yu, WC; Guo, F; Qian, YT. Control to synthesize Bi2S3 nanowires by a simple inorganic-surfactant-assisted solvothermal process. *Nanotechnol.*, 2005, 16, 1771-1775.

[256] Wang, Q; Jiang, CL; Yu, CF; Chen, QW. General solution-based route to V-VI semiconductors nanorods from hydrolysate. *J. Nanoparticle Res.,* 2007, 9, 269-274.

[257] Fan, LB; Song, HW; Zhao, HF; Pan, GH; Yu, HQ; Bai, X; Li, SW; Lei, YQ; Dai, QL; Qin, RF; Wang, T; Dong, B; Zheng, ZH; Ren, XG. Solvothermal synthesis and photoluminescent properties of ZnS/Cyclohexylamine: inorganic-organic hybrid semiconductor nanowires. *J. Phys. Chem. B,* 2006, 110, 12948 -12953.

[258] Li, YD; Liao, HW; Ding, Y; Fan, Y; Zhang, Y; Qian, YT. Solvothermal Elemental Direct Reaction to CdE (E = S, Se, Te) Semiconductor Nanorod. *Inorg. Chem.,* 1999, 38, 1382 -1387.

[259] Liu, Y; Qiu, HY; Xu, Y; Wu, D; Li, MJ; Jiang, JX; Lai, GQ. Selective Synthesis of Wurtzite CdSe Nanorods and Zinc Blend CdSe Nanocrystals through a Convenient Solvothermal Route. *J. Nanoparticle Res.,* 2007, 9, 745-752.

[260] Thongtem, T; Phuruangrat, A; Thongtem, S. Free surfactant synthesis of microcrystalline CdS by solvothermal reaction. *Mater. Lett.,* 2007, 61, 3235-3238.

[261] Ma, C; Moore, D; Ding, Y; Li, J; Wang, ZL. Nanobelt and nanosaw structures of II-VI semiconductors. *Int. J. Nanotechnol.,* 2004, 4, 431-451.

[262] Yao, WT; Yu, SH; Wu, QS. From mesostructured wurtzite ZnS-Nanowire/Amine nanocomposites to ZnS nanowires exhibiting quantum size effects: A mild-solution chemistry approach. *Adv. Funct. Mater.,* 2007, 17, 623–631.

[263] Nath, M; Parkinson, BA. A simple sol–gel synthesis of superconducting MgB_2 nanowires. *Adv. Mater.,* 2006, 18, 1865–1868.

[264] Jia, FL; Zhang, LZ; Shang, XY; Yang, Y. Non-aqueous sol–gel approach towards the controllable synthesis of Nickel nanospheres, nanowires, and nanoflowers. *Adv. Mater.* 2008, *20,* 1050–1054.

[265] Woo, K; Lee, HJ; Ahn, JP; Park, YS. Sol-Gel mediated synthesis of Fe_2O_3 nanorods. *Adv. Mater.,* 2003, 15, 1761-1764.

[266] Yang, Q; Sha, J; Ma, XY; Yang, DR. Synthesis of NiO nanowires by a sol-gel process. *Mater. Lett.,* 2005, 59, 1967-1970.

[267] Kuiry, SC; Megen, E; Patil, SD; Deshpande, SA; Seal, S; Solution-Based Chemical Synthesis of Boehmite Nanofibers and Alumina Nanorods. *J. Phys. Chem. B,* 2005, 109, 3868-3872.

[268] Hou, YD; Hou, L; Zhang, TT; Zhu, MK; Wang, H; Yan, H. $(Na_{0.8}K_{0.2})$ $(0.5)Bi_{0.5}TiO_3$ nanowires: Low-temperature sol-gel-hydrothermal synthesis and densification. *J. Am. Ceram. Soc.,* 2007, 90, 1738-1743.

[269] Hou, YD; Hou, L; Zhu, MK; Yan, H. Synthesis of $(K_{0.5}Bi_{0.5})$ $(0.4)Ba_{0.6}TiO_3$ nanowires and ceramics by sol-gel-hydrothermal method. *Appl. Phys. Lett.,* 2006, 89, Art. No. 243114.

[270] Hou, L; Hou, YD; Song, XM; Zhu, MK; Wang, H; Yan, H. Sol-gel-hydrothermal synthesis and sintering of $K_{0.5}Bi_{0.5}TiO_3$ nanowires. *Mater. Res. Bull.,* 2006, 41, 1330-1336.

[271] Xu, HY; Wei, SQ; Wang, H; Zhu, MK; Yu, R; Yan, H. Preparation of shape controlled SrTiO3 crystallites by sol-gel-hydrothermal method. *J. Cryst. Grow.,* 2006, 292, 159-164.

[272] Wu, GS; Xie, T; Yuan, XY; Li, Y; Yang, L; Xiao, YH; Zhang, LD. Controlled synthesis of ZnO nanowires or nanotubes via sol-gel template process. *Solid State Commun.*, 2005, 134, 485-489.

[273] Cao, HQ; Qiu, XQ; Yu, LA; Zhao, MJ; Zhu, QM. Sol-gel synthesis and photoluminescence of p-type semiconductor Cr_2O_3 nanowires. *Appl. Phys. Lett.,* 2006, Art. No. 241112.

[274] Yang, Z; Huang, Y; Dong, B; Li, HL; Shi, SQ. Sol-gel template synthesis and characterization of $LaCoO_3$ nanowires. *Appl. Phys. A-Mater. Sci. Process.* 2006, 84, 117-122.
[275] Yang, Z; Huang, Y; Dong, B; Li, HL. Controlled synthesis of highly ordered $LaFeO_3$ nanowires using a citrate-based sol-gel route. *Mater. Res. Bull.,* 2006, 41, 274-281.
[276] Yang, Z; Huang, Y; Dong, B; Li, HL. Fabrication and structural properties of LaFeO3 nanowires by an ethanol-ammonia-based sol-gel template route. *Appl. Phys. A- Mater. Sci. Process.,* 2005, 81, 453-457.
[277] Yang, Z; Huang, Y; Dong, B; Li, HL. Template induced sol-gel synthesis of highly ordered $LaNiO_3$ nanowires. *J. Solid State Chem.,* 2005, 178, 1157-1164.
[278] Ma, XY; Zhang, H; Xu, J; Niu, JJ; Yang, Q; Sha, JA; Yang, DR. Synthesis of $La_{1-x}Ca_xMnO_3$ nanowires by a sol-gel process. *Chem. Phys. Lett.,* 2002, 363, 579-582.
[279] Yang, Z; Huang, Y; Dong, B; Li, HL; Shi, SQ. Densely packed single-crystal $Bi_2Fe_4O_9$ nanowires fabricated from a template-induced sol-gel route. *J. Solid State Chem.,* 2006, 179, 3324-3329.
[280] Liu, XH; Wang, JQ; Zhang, JY; Yang, SR. Fabrication and characterization of Zr and Co co-doped $LiMn_2O_4$ nanowires using sol-gel-AAO template process. *J. Mater. Sci. –Mater. Electr.,* 2006, 17, 865-870.
[281] McCann, JT; Li , D; Xia, YN. Electrospinning of nanofibers with core-sheath, hollow, or porous structures. *Mater. Chem.,* 2005, 15, 735 – 738.
[282] Yuh, J; Perez, L; Sigmund, WM; Nino, JC. Sol-gel based synthesis of complex oxide nanofibers. *J. Sol-gel Sci. Tech.,* 2007, 42, 323-329.
[283] Zhan, SH; Gong, CR; Chen, DR; Jiao, XL. Preparation of $ZnFe_2O_4$ nanofibers by sol-gel related electrospinning method. *J. Disper. Sci. Tech.,* 2006, 27, 931-933.
[284] Zaidi, SSA; Rohani, S. Progress towards a dry process for the synthesis of zeolite - A review. *Rev. Chem. Eng.,* 2005, 21, 265-306.
[285] Hidai, H; Tokura, H. Hydrothermal-reaction-assisted laser machining of cubic boron nitride. *J. Am. Ceram. Soc.,* 2006, 89, 1621-1623.
[286] Shen, SC; Ng, WK; Chen, Q; Zeng, XT; Chew, MZ; Tan, RBH. Solid-Phase Low Temperature Steam-Assisted Synthesis of Thermal Stable Alumina Nanowires. *J. Nanosci. Nanotechnol.* 2007, 7, 2726-2733.
[287] Popa, AF; Rossignol, S; Kappenstein, C. Ordered structure and preferred orientation of boehmite films prepared by the sol-gel method. *J. Non-Cryst. Solids,* 2002, 306, 169-174.

[288] Wilson, SJ; McConnell, JDC. A kinetic study of the system boehmite/alumina (.gamma.-AlOOH/Al_2O_3). *J. Solid State Chem.,* 1980 34, 315-322.

[289] Pierre, AC; Uhlmann, DR. Amorphous aluminum hydroxide gels. *J. Non-Cryst. Solid*, 1986 82, 271-276.

[290] Shen, SC; Chen, Q; Chow, PS; Tan, GH, Zeng, XT; Wang, Z; Tan, RBH. Steam-assisted solid wet-gel synthesis of high-quality nanorods of boehmite and alumina, *J. Phys. Chem. C,* 2007, 111, 700-707.

[291] Pozarnsky, GA; McCormick, AV. Multinuclear NMR-study of aluminosilicate sol-gel synthesis using the prehydrolysis method. *J. Non-Cryst. Solid*, 1995, 190, 212-225.

[292] Bagshaw, SA; Pinnavaia, T J. Mesoporous alumina molecular sieves. *Angew. Chem. Int. Ed. Eng.,* 1996, 35, 1102-1105.

[293] Chen, FR; Davis, JG; Fripiat, JJ. Aluminum coordination and lewis acidity in transition alumas. *J. Catal.*, 1992, 133, 263-278.

[294] Baraton, M I; Quintard, P. Infrared evidence of order-disorder phase transitions (gamma→delta→ alpha) in alumina. *J. Mol. Struct.* 1982, 79, 337-340.

[295] Tsyganenko, AA; Mardilovich, PP. Structure of alumina surfaces. *J. Chem. Soc., Faraday Trans.,* 1996, 92, 4843-4852.

[296] Shen, SC; Ng, WK; Chen, Q; Zeng, XT; Tan, RBH. Novel synthesis of lace-like nanoribbons of boehmite and γ-alumina by dry gel conversion method. *Mater. Lett.,* 2007, 61, 4280–4282.

[297] Jin CF, Yuan X, Ge WW, Hong JM, Xin XQ. Synthesis of ZnO nanorods by solid state reaction at room temperature, *Nanotechnol.*, 2003, 14, 667-669.

[298] Buscaglia, MT; Harnagea, C; Dapiaggi, M; Buscaglia, V; Pignolet, A; Nanni, P. Ferroelectric BaTiO3 Nanowires by a Topochemical Solid-State Reaction. *Chem Mater.,* 2009, 21, 5058–5065.

[299] Du, XS; Zhou, CF; Wang, GT; Mai, WY. Novel solid-state and template-free synthesis of branched polyaniline nanofibers. *Chem. Mater.,* 2008, 20, 3806–3808.

[300] Huang, ZM; Zhang, YZ; Kotaki, M; Ramakrishna, S. Review on polymer nanofibers by electrospinning and their applications in nanocomposites. *Composites Sci. Tech.*, 2003, 63, 2223-2253.

[301] Chronakis, IS. Novel nanocomposites and nanoceramics based on polymer nanofibers using electrospinning process - A review. *J. Mater. Process. Tech.,* 2005, 167, 283-293.

[302] Pham, QP; Sharma, U; Mikos, AG. Electrospinning of polymeric nanofibers for tissue engineering applications: A review. *Tissue Eng.,* 2006, 12, 1197-1121.

[303] Li, D; Babel, A; Jenekhe, SA; Xia, YN. Nanofibers of conjugated polymers prepared by electrospinning with a two-capillary spinneret. *Adv. Mater.,* 2004, 16, 2062- 2066.

[304] Katta, P; Alessandro, M; Ramsier, RD; Chase, GG. Continuous electrospinning of aligned polymer nanofibers onto a wire drum collector. *Nano Lett.,* 2004, 4, 2215-2218.

[305] Lv, YY; Wu, J; Wan, LS; Xu, ZK. Novel porphyrinated polyimide nanofibers by electrospinning. *J. Phys. Chem. C,* 2008, *112,* 10609–10615.

[306] Zhang, W; Yan, EY; Huang, ZH; Wang, C; Xin, Y; Zhao, Q; Tong, YB. Preparation and study of PPV/PVA nanofibers via electrospinning PPV precursor alcohol solution. *Euro. Polym. J.,* 2007, 43, 802-807.

[307] Reneker, DH; Yarin, AL; Zussman, E; Xu, H. Electrospinning of nanofibers from polymer solutions and melts. *Advance in Appl. Mechanics,* 2007, 41, 43-195.

[308] Talwar, S; Hinestroza, J; Pourdeyhimi, B; Khan, SA. Associative polymer facilitated electrospinning of nanofibers. *Macromolecules,* 2008, 41, 4275-4283.

[309] Sun, XY; Shankar, R; Borner, HG; Ghosh, TK; Spontak, RJ. Field-driven biofunctionalization of polymer fiber surfaces during electrospinning. *Adv. Mater.,* 2007, 19, 87-91.

[310] Diaz, JE; Barrero, A; Marquez, M; Loscertales, IG. Controlled encapsulation of hydrophobic liquids in hydrophilic polymer nanofibers by co-electrospinning. *Adv. Funct. Mater.,* 2006, 16, 2110-2116.

[311] Spasova, M; Stoilova, O; Manolova, N; Rashkov, I; Altankov, G. Preparation of PLIA/PEG nanofibers by electrospinning and potential applications. *J. Bioactive Compat. Polym.,* 2007, 22, 62-76.

[312] Zhang, JF; Yang, DZ; Xu, F; Zhang, ZP; Yin, RX; Nie, J. Electrospun core-shell structure nanofibers from homogeneous solution of poly(ethylene oxide)/chitosan. *Macromolecules*, 2009, 42, 5278–5284.

[313] Moran-Mirabal, J; Slinker, JD; DeFranco, JA; Verbridge, SS; Ilic, R; Flores-Torres,S; Abruna, H; Malliaras, GG; Craighead, HG. Electrospun light-emitting nanofibers. , *Nano Lett.*, 2007, 7, 458-463.

[314] Wang, HY; Lu, XF; Zhao, YY; Wang, C. Preparation and characterization of ZnS : Cu/PVA composite nanofibers via electrospinning. *Mater. Lett.,* 2006, 60, 2480-2484.

[315] Faridi-Majidi, R; Sharifi-Sanjani, N. In situ synthesis of iron oxide nanoparticles on poly(ethylene oxide) nanofibers through an electrospinning process. *J. Appl. Polym. Sci.,* 2007, 105, 1351-1355.

[316] Jeong, JS; Jeon, SY; Lee, TY; Park, JH; Shin, JH; Alegaonkar, PS; Berdinsky, AS; Yoo, JB. Fabrication of MWNTs/nylon conductive composite nanofibers by electrospinning. *Diamond Related Mater.,* 2006, 15, 1839-1843.

[317] Chen, LJ; Liao, JD; Lin, SJ; Chuang, YJ; Fu, YS. Synthesis and characterization of PVB/silica nanofibers by electrospinning process. *Polymer* , 2009, 50, 3516–3521.

[318] Kedem, S; Rozen, D; Cohen, Y; Paz, Y. Enhanced stability effect in composite polymeric nanofibers containing titanium dioxide and carbon nanotubes. *J. Phys. Chem. C* 2009, *113,* 14893–14899.

[319] Panda, PK; Ramakrishna, S. Electrospinning of alumina nanofibers using different precursors. *J. Mater. Sci.,* 2007, 42, 2189-2193.

[320] Patel, AC; Li, SX; Wang, C; Zhang, WJ; Wei, Y. Electrospinning of porous silica nanofibers containing silver nanoparticles for catalytic applications. *Chem. Mater.,* 2007, 19, 1231-1238.

[321] Zhang, YF; Li, JY; Li, Q; Zhu, L; Liu, XD; Zhong, XH; Meng, J; Cao, XQ. Preparation of In_2O_3 ceramic nanofibers by electrospinning and their optical properties. *Scripta Mater.,* 2007, 56, 409-412.

[322] Li, SZ; Shao, CL; Liu, YC; Tang, SS; Mu, RX. Nanofibers and nanoplatelets of MoO_3 via an electrospinning technique. *J. Phys. Phys. Chem. Solid.,* 2006, 67, 1869-1872.

[323] Ostermann, R; Li, D; Yin, YD; McCann, JT; Xia, YN. V_2O_5 nanorods on TiO_2 nanofibers: A new class of hierarchical nanostructures enabled by electrospinning and calcination. *Nano Lett.,* 2006, 6, 1297-1302.

[324] Widiyandari, H; Munir, HM; Iskandar, F; Okuyama, K . Morphology-controlled synthesis of chromia–titania nanofibers via electrospinning followed by annealing. *Mater. Chem. Phys.,* 2009, 116, 169–174.

[325] Yuh, J; Perez, L; Sigmund, WM; Nino, JC. Electrospinning of complex oxide nanofibers. *Phys. E-low-dimensional Sys. Nanostruct.,* 2007, 37, 254-259.

[326] Hou,ZY; Chai, RT; Zhang, ML; Zhang, CM; Chong, P; Xu, ZH; Li, GG; Lin, J. Fabrication and luminescence properties of one-dimensional $CaMoO_4$: Ln_3 (Ln = Eu, Tb, Dy) nanofibers via electrospinning process. *Langmuir* 2009, 25(20), 12340–12348.

[327] Tang, MH; Shu, W; Yang, F; Zhang, J; Dong, GJ; Hou, JW. The fabrication of La-substituted bismuth titanate nanofibers by electrospinning. Nanotechnol. 2009, 20, Art 385602 (4pp).

[328] Wu, H; Sun, Y; Lin, DD; Zhang, R; Zhang, C; Pan, W. GaN Nanofibers based on electrospinning: facile synthesis, controlled assembly, precise doping, and application as high performance UV photodetector. *Adv. Mater.,* 2009, 21, 227–231.

[329] Qiu, YJ; Yu, J; Rafique, J; Yin, J; Bai, XD; Wang, E. Large-scale production of aligned long boron nitride nanofibers by multijet/multicollector. Electrospinning . *J. Phys. Chem. C,* 2009, 113, 11228–11234.

[330] Shui, JL; Li, JCM. Platinum nanowires produced by electrospinning. *Nano Lett.* 2009, 9, 1307-1314.

[331] Li, D; Xia, YN. Direct Fabrication of Composite and Ceramic Hollow Nanofibers by Electrospinning. *Nano Lett.* 2004, 4, 933-938.

[332] Hassenkam, T; Norgaard, K; Iversen, L; Kiely, CJ; Brust, M; Bjornholm, T. Fabrication of 2D gold nanowires by self-assembly of gold nanoparticles on water surfaces in the presence of surfactants. *Adv. Mater.,* 2002, 14, 1126-1130.

[333] Zhang, H; Yang, DR; Ma, XY; Ji, YJ; Li, SZ; Que, DL. Self-assembly of CdS: from nanoparticles to nanorods and arrayed nanorod bundles. *Mater. Chem. Phys.,* 2005, 93, 65-69.

[334] Zhang, CY; Zhang, XJ; Zhang, XH; Fan, X; Jie, JS; Chang, JC; Lee, CS; Zhang, WJ; Lee, ST. Facile one-step growth and patterning of aligned squaraine nanowires via evaporation-induced self-assembly. *Adv. Mater.,* 2008, 20, 1716–1720.

[335] Feng, JT; Yan, W; Zhang, LZ. Synthesis of polypyrrole micro/nanofibers via a self-assembly process. *Microchim Acta,* 2009, 166, 261–267.

[336] Cui, H; Muraoka, T; Cheetham, AG; Stupp, SI. Self-Assembly of giant peptide nanobelts. *Nano Lett.,* 2009, 9, 945-951.

[337] Tang, ZY; Kotov, NA; Giersig, M. Spontaneous organization of single CdTe nanoparticles into luminescent nanowires. *Science,* 2002, 297, 237-240.

[338] Tang, ZY; Kotov, NA. One-dimensional assemblies of nanoparticles: preparation, properties and promise. *Adv. Mater.,* 2005, 17, 951-962.

[339] Kuiry, SC; Patil, SD; Deshpande, S; Seal, S. Spontaneous self-assembly of cerium oxide nanoparticles to nanorods through supraaggregate formation. *J. Phys. Chem. B,* 2005, 109, 6936-6939.

[340] Levine, LE; Long, GG; Ilavsky, J; Gerhardt, RA; Ou, R; Parker, CA. Self-assembly of carbon black into nanowires that form a conductive three dimensional micronetwork. *Appl. Phys. Lett.,* 2007, 90, Art. No. 014101.

[341] Schoiswohl, J; Mittendorfer, F; Surnev, S; Ramsey, MG; Andersen, JN; Netzer, FP. The self-assembly of metallic nanowires. *Surf. Sci.,* 2006, 600, L274-L280.

[342] Zhang, ZL; Tang, ZY; Kotov, NA; Glotzer, SC. Simulations and analysis of self-assembly of CdTe nanoparticles into wires and sheets. *Nano Lett.*, 2007, 7, 1670-1675.

[343] Deng, ZT; Chen, D; Tang, FQ; Meng, XW; Ren, J; Zhang, L. Orientated attachment assisted self-assembly of Sb2O3 nanorods and nanowires: End-to-end versus side-by-side. *J. Phys. Chem. C,* 2007, 111, 5325-5330.

[344] Sun, SH; Yang, DQ; Villers, D; Zhang, GX; Sacher, E; Dodelet, JP. Template- and surfactant-free room temperature synthesis of self-assembled 3D Pt nanoflowers from single-crystal nanowires. *Adv. Mater.,* 2008, 20, 571–574

[345] Feng, XL; Liang, YY; Zhi, LJ; Thomas, A; Wu, DQ; Lieberwirth, I; Kolb, U; Müllen, K. Synthesis of microporous carbon nanofibers and nanotubes from conjugated polymer network and evaluation in electrochemical capacitor. *Adv. Funct. Mater.,* 2009, 19, 2125–2129.

[346] Sheparovych, R; Sahoo, Y; Motornov, M; Wang, SM; Luo, H; Prasad, PN; Sokolov, I; Minko, S. Polyelectrolyte stabilized nanowires from Fe_3O_4 nanoparticles via magnetic field induced self-assembly. *Chem. Mater.,* 2006, 18, 591-593.

[347] Jia, YS; Chen, QW; Wu, MZ. Room temperature self-assembly growth of cobalt nanowires under magnetic fields. *Int. J Morden Phys.,* 2005, 19, 2728-2733.

[348] Wu, LY; Lian, JB; Sun, GX; Kong, XR; Zheng, WJ. Synthesis of zinc hydroxyfluoride nanofibers through an ionic liquid assisted microwave irradiation method. *Eur. J. Inorg. Chem.* , 2009, 2897–2900.

[349] Zou, GF; Zhang, DW; Dong, C; Li, H; Xiong, K; Fei, LF; Qian, YT. Carbon nanofibers: Synthesis, characterization, and electrochemical properties. *Carbon*, 2006, 44, 828–832.

[350] Pesant, L; Wine, G; Vieira, R; Leroi, P; Keller, N; Phamhuu, C; Ledoux, MJ. Large-scale synthesis of carbon nanofibers by catalytic decomposition of hydrocarbon. *Stud. Surf. Sci. Catal.,* 2002, 143, 193-200.

[351] Segura, R; Tello, A; Cardenas, G; Häberle, P. Synthesis of carbon nanotubes and nanofibers by decomposition of acetylene over a SMAD palladium catalyst. *Phys. Stat. Sol.,* 2007, 204, 513– 517.

[352] Tanemura, M; Okita, T; Tanaka, J; Kitazawa, M; Itoh, K; Miao, L; Tanemura, S; Lau, Sp; Yang, HY; Huang, L. Room-temperature growth and application of carbon nanofibers: a review. *IEEE. Trans. Nanotechnol.*, 2006, 5, 587-594.

[353] M. Minea, TM; Point, S; Granier, A; Touzeau, M. Room temperature synthesis of carbon nanofibers containing nitrogen by plasma-enhanced chemical vapor deposition. *Appl. Phys. Lett.,* 2004, 85, 1244-1246.

[354] Mozetic, M. Cvlbar, U; Sunkara, MK; Vaddiraju, S. A method for the rapid synthesis of large quantities of metal oxide nanowires at low temperatures. *Adv. Mater.,* 2005, 17, 2183-2142.

[355] Zhang, HG; Wang, XH; Li, J; Wang, FS. Facile synthesis of polyaniline nanofibers using pseudo-high dilution technique. *Synthetic Metals,* 2009, 159, 1508–1511.

[356] Kumar, S; Singh, V; Aggarwal, S; Mandal, UK . Synthesis of 1-dimensional polyaniline nanofibers by reverse microemulsion. *Colloid Polym Sci.,* 2009, 287:1107–1110.

[357] Zhou, CF; Du, XS; Liu, Z; Ringer, SP; Mai, YW. Solid phase mechanochemical synthesis of polyaniline branched nanofibers. *Synthetic Metals,* 2009, 159. 1302-1307.

[358] Weitz, RT; Harnau, L; Rauschenbach, S; Burghard, M; Kern, K. Polymer nanofibers via nozzle-free centrifugal spinning. Nano Lett. 2008, 8, 1187-1191.

[359] Teo, BK; Sun, XH. Silicon-Based Low-Dimensional Nanomaterials and Nanodevices. *Chem. Rev.,* 2007, 107, 1454-1532.

[360] Hu, J; Odom, TW; Lieber, CM. Chemistry and physics in one dimension: Synthesis and properties of nanowires and nanotubes. *Acc. Chem. Res.* 1999, *32*, 435.

[361] Bethoux, JM; Happy, H; Dambrine, G; Derycke, V; Goffman, M; Bourgoin, JP. An 8-GHz f(t) carbon nanotube field-effect transistor for gigahertz range applications. *IEEE Electron Device Lett.,* 2006, 27, 681-683.

[362] Fu, L; Liu, YQ; Liu, ZM; Han, BX; Cao, LC; Wei, DC; Yu, G; Zhu, DB. Carbon nanotubes coated with alumina as gate dielectrics of field-effect transistors. *Adv. Mater.,* 2006, *18*, 181–185.

[363] Pesetski, AA; Baumgardner, JE; Folk, E; Przybysz, JX; Adam, JD; Zhang, H. Carbon nanotube field-effect transistor operation at microwave frequencies. *Appl. Phys. Lett.,* 2006, 88, Art No. 113103.

[364] Wu, F; Tsuneta, T; Tarkiainen, R; Gunnarsson, D; Wang, TH; Hakonen, PJ. Shot noise of a multiwalled carbon nanotube field effect transistor. *Phys. Rev.,* 2007,75, Art No.125419.

[365] Park, JY. Carbon nanotube field-effect transistor with a carbon nanotube gate electrode. *Nanotechnol.* 2007, 18, Art. No. 095202.

[366] Zhang, M; Huo, X; Chan, PCH; Liang, Q; Tang, ZK. Radio-frequency transmission properties of carbon nanotubes in a field-effect transistor configuration. *IEEE Electron Device Lett.,* 2006, 27, 688-670.

[367] Chen, BH; Wei, JH; Lo, PY; Wang, HH; Lai, MJ; Tsai, MJ; Chao, TS; Lin, HC; Huang, TY. A carbon nanotube field effect transistor with tunable conduction-type by electrostatic effects. *Sol. Stat. Electronics,* 2006, 50, 1341–1348.

[368] Liu, F; Wang, KL; Zhang, DH; Zhou, CW. Noise in carbon nanotube field effect transistor. *Appl. Phys. Lett.,* 2006, 89, Art No. 063116.

[369] Raychowdhury, A; Keshavarzi, A; Kurtin, J; De, V; Roy, K. Carbon nanotube field-effect transistors for high-performance digital circuits—DC Analysis and modeling toward optimum transistor structure. *IEEE Trans. Electron Device*, 2006, 53, 2711-2717.

[370] Ishibashi, K; Moriyama, S; Tsuya, D; Fuse, T. Suzuki, M. Quantum-dot nanodevices with carbon nanotubes. *J. Vacuum Sci. Tech.,* 2006, 24, 1349-1355.

[371] Ma, RM; Dai, L; Qin, GG. Enhancement-mode metal-semiconductor field-effect transistors based on single *n*-CdS nanowires. *Appl. Phys. Lett.,* 2009, 90, Art No. 093109.

[372] Koo, SM; Edelstein, MD; Li, QL; Richter, CA; Vogel, EM. Silicon nanowires as enhancement-mode Schottky barrier field-effect transistors. *Nanotechnol.,* 2005, 16, 1482-1485.

[373] Zheng, GF; Lu, W; Jin, S; Leiber, CM. Synthesis and fabrication of high-performance n-type silicon nanowire transistors. *Adv. Mater.,* 2004, 16, 1890-1893.

[374] Arnold, MS; Avouris, P; Pan, ZW; Wang, ZL. Field-effect transistors based on single semiconducting oxide nanobelts. *J. Phys. Chem. B,* 2003, 107, 659-633.

[375] Umar, A; Kim, BK; Kim, JJ; Hahn, YB. Optical and electrical properties of ZnO nanowires grown on aluminium foil by non-catalytic thermal evaporation. *Nanotechnol.*, 2007, 18, Art. No. 175606.

[376] Keem, K; Jeong, DY; Kim, S; Lee, MS; Yeo, IS; Chung, UI; Moon, JT. Fabrication and device characterization of omega-shaped-gate ZnO nanowire field-effect transistors. *Nano Lett.,* 2006, 6, 1454-1458.

[377] Basu, D; Wang, L; Dunn, L; Yoo, B; Nadkarni, S; Dodabalapur, A; Heeney, M; McCulloch, I. Direct measurement of carrier drift velocity and mobility in a polymer field-effect transistor. *Appl. Phys. Lett.*, 2006, 89, Art No. 242104.

[378] Wanekaya, AK; Bangar, MA; Yun, M; Chen, W; Myung, NV; Mulchandani, A. Field-effect transistors based on single nanowires of conducting polymers. *J. Phys. Chem. C,* 2007, *111,* 5218-5221.

[379] Zhitenew, NB; Sidorenko, A; Tennant, DM; Cirelli, RA. Chemical modification of the electronic conducting states in polymer nanodevices. *Nature Nanotechnol.,* 2007, 2, 237-242.

[380] Berdichevsky, Y; Lo, YH. Polypyrrole Nanowire actuators. *Adv. Mater.,* 2006, 18, 122-125.

[381] Zhong, ZH; Qian, F; Wang, DL; Lieber, CM. Synthesis of p-type gallium nitride nanowires for electronic and photonic nanodevices. *Nano Lett.,* 2003, 3, 343-346.

[382] He, JH; Hsin, CL; Liu, J; Chen, LJ; Wang, ZL. Piezoelectric Gated Diode of a Single ZnO Nanowire. *Adv. Mater.,* 2007, *19*, 781–784

[383] Choi, MY; Choi, D; Jin, MJ; Kim, I; Kim, SH; Choi, JY; Lee, SY; Kim, JM; Kim, SW. Mechanically powered transparent flexible charge-generating nanodevices with piezoelectric ZnO nanorods. *Adv. Mater.,* 2009, 21, 2185–2189.

[384] Gao, PX; Song, JH; Liu, J; Wang, ZL. Nanowire piezoelectric nanogenerators on plastic substrates as flexible power sources for nanodevices. *Adv. Mater.,* 2007, *19*, 67–72.

[385] Alexe, M; Senz, S; Schubert, MA; Hesse, D; Güsele, U. Energy harvesting using nanowires?. *Adv. Mater.,* 2008, 20, 4021–4026.

[386] Wang, ZL. Energy Harvesting using piezoelectric nanowires–a correspondence on ''energy harvesting using nanowires?'' by Alexe *et al. Adv. Mater.,* 2009, 21, 1311–1315.

[387] Zhong, ZH; Wang, DL; Cui, Y; Bockrath, MW; Lieber, CM. Nanowire Crossbar Arrays as Address Decoders for Integrated Nanosystems. *Science*, 2003, 302, 1377-1379.

[388] Service, RF. Molecular electronics: nanodevices make fresh strides toward reality, *Science*, 2003, 302, 1310.

[389] Park, WI; Kim, JS; Yi, GC; Lee, HJ. ZnO nanorod logic circuits. *Adv.Mater.,* 2005, 17, 1393-1397.

[390] Willner, I; Basnar, B; Willner, B. From molecular machines to microscale motility of objects: application as "smart materials", sensors, and nanodevices. *Adv. Funct. Mater.,* 2007, *17*, 702–717

[391] Bakewell, DJG; Nicolau, DV. Protein linear molecular motor-powered nanodevices. *Australian J. Chem.,* 2007, 60, 314-332.

[392] Liedl, T; Sobey, TL; Simmel, FC. DNA-based nanodevices. *Nano Today*, 2007, 2, 36-41.

[393] Martin, C; Grisolia, J; Ressier, L; Respaud, M; Peyrade, JP; Carcenac, F; Vieu, C. Fabrication of nanodevices for magneto-transport measurements through nanoparticles. *Microelectro. Eng.,* 2004, 73-4, 627-631.

[394] Cho, A. Nanotechnology - Pretty as you please, curling films turn themselves into nanodevices. *Science,* 2006, 313, 164-165.

[395] Gao, ZQ; Agarwal, A; Trigg, AD; Singh, N; Fang, C; Tung, CH; Fan, Y; Buddharaju, KD; Kong, JM. Silicon nanowire arrays for label-free detection of DNA. *Anal. Chem*., 2007, 79, 3291-3297.

[396] Yang, K; Wang, H; Zou, K; Zhang, XH. Gold nanoparticle modified silicon nanowires as biosensors. *Nanotechnol*., 2006, 17, S276-S279.

[397] Lu, YS; Yang, MH; Qu, FL; Shen, GL; Yu, RQ. Amperometric biosensors based on platinum nanowires. *Anal. Lett.,* 2007, 40, 875-886.

[398] Lu, XB; Wen, ZH; Li, JH. Hydroxyl-containing antimony oxide bromide nanorods combined with chitosan for biosensors. *Biomater.,* 2006, 27, 5740-5747.

[399] Byun, KM; Yoon, SJ; Kim, D; Kim, SJ. Experimental study of sensitivity enhancement in surface plasmon resonance biosensors by use of periodic metallic nanowires. *Optics Lett.,* 2007, 32, 1902-1904.

[400] Feng, CL; Zhong, XH; Steinhart, M; Caminade, AM; Majoral, JP; Knoll, W. Graded-Bandgap Quantum- Dot-Modified Nanotubes: A Sensitive Biosensor for Enhanced Detection of DNA Hybridization. *Adv. Mater.*, 2007, 19, 1933-1936.

[401] Shao, MW; Shan, YY; Wong, NB; Lee, ST. Silicon nanowire sensor for bioanalytical application: glucose and hydrogen peroxide detection. *Adv. Funct. Mater.,* 2005, 15, 1478-1482.

[402] Murphy, CJ. Biosensors plasmons spring into action. *Nature Mater.,* 2007, 6, 259-260.

[403] Allen, BL; Kichambare, PD; Star, A. Carbon nanotube field-effect-transistor-based biosensors. *Adv. Mater.,* 2007, 19, 1439–1451

[404] Nandanan, E; Jana, NR; Ying, JY. Functionalization of gold nanospheres and nanorods by Cchitosan oligosaccharide derivatives. *Adv. Mater.,* 2008, 20, 2068–2073.

[405] Atashbar, MZ; Singamaneni, S. Room temperature gas sensor based on metallic nanowires. *Sensors and Actuators B,* 2005, 111-112, 13-21.

[406] Favier, F; Walter, EC; Zach, MP; Benter, T; Penner, RM. Hydrogen sensors and switches from electrodeposited palladium mesowire arrays. *Science,* 2001, 293, 2227-2231.

[407] Kim, KT; Sim, SJ; Cho, SM. Hydrogen Gas Sensor Using Pd Nanowires Electro-Deposited Into Anodized Alumina Template. *IEEE Sensor J.,* 2006, 6, 509-513.

[408] Rout, CS; Krishna, SH; Vivekchand, SRC; Govindaraj, A; Rao, CNR. Hydrogen and ethanol sensors based on ZnO nanorods, nanowires and nanotubes. *Chem. Phys. Lett.,* 2006, 418-586-590.

[409] Rout, CS; Raju, AR; Govindaraj, A; Rao, CNR. Ethanol and hydrogen sensors based on ZnO nanoparticles and nanowires. *J. NanoSci. Nanotechnol.,* 2007, 7, 1923-1929.

[410] Chen, J; Xu, LN; Li, WY; Gou, XL. α-Fe_2O_3 nanotubes in gas sensor and lithium-ion battery application. *Adv. Mater.,* 2005, 17, 582-586.

[411] Rout, CS; Krishna, SH; Vivekchand, SRC; Govindaraj, A; Rao, CNR. Hydrogen and ethanol sensors based on ZnO nanorods, nanowires and nanotubes. *Chem. Phys. Lett.,* 2006, 418, 586-590.

[412] Fowler, JD; Virji, S; Kaner, RB; Weiller, BH. Hydrogen detection by polyaniline nanofibers on gold and platinum electrodes. *J. Phys. Chem. C,* 2009, 113, 6444–6449.

[413] Ding, DY; Chen, Z; Rajaputra, S; Singh, V. Hydrogen sensors based on aligned carbon nanotubes in an anodic aluminum oxide template with palladium as a top electrode. *Sensors Actuators B: Chem.,* 2007, 124, 12-17.

[414] Rout, CS; Kulkarni, GU; Rao, CNR. Room temperature hydrogen and hydrocarbon sensors based on single nanowires of metal oxides. *J. Phys. D-Appl. Phys.,* 2007, 40, 2777-2782.

[415] Si, SF; Li, CH; Wang, X; Peng, Q; Li, YD. Fe_2O_3/ZnO core-shell nanorods for gas sensors. *Sensor Actuator B-Chem.,* 2006, 119, 52-56.

[416] Wu, CZ; Yin, P; Zhu, X; OuYang, CZ; Xie, Y. Synthesis of hematite (alpha-Fe_2O_3) nanorods: Diameter-size and shape effects on their applications in magnetism, lithium ion battery, and gas sensors. *J. Phys. Chem. B,* 2006, 110, 17806-17812.

[417] Dua,V; Surwade, SP; Ammu, S; Zhang, XY; Jain, S; Manohar, SK. Chemical vapor detection using parent polythiophene nanofibers. *Macromolecules*, 2009, 42, 5414–5415.

[418] Wang, CH; Chu, XF; Wu, MW. Detection of H_2S down to ppb levels at room temperature using sensors based on ZnO nanorods. *Sensor Actuator B-Chem.,* 2006, 113, 320-323.

[419] Song, XF; Wang, ZJ; Liu, YB; Wang, C; Li, LJ. A highly sensitive ethanol sensor based on mesoporous ZnO-SnO_2 nanofibers. Nanotechnol. 2009, 20, Art No. 075501.

[420] Rout, CS; Ganesh, K; Govindaraj, A; Rao, CNR. Sensors for the nitrogen oxides, NO2, NO and N_2O, based on In_2O_3 and WO_3 nanowires. *Appl. Phys. A-Mater. Sci. Procss.,* 2006, 85, 241-246.

[421] Mcalpine, MC; Ahmad, H; Wang, DW; HEATH, JR. Highly ordered nanowire arrays on plastic substrates for ultrasensitive flexible chemical sensors. *Nature Mater.,* 2007, 6, 379-384.

[422] [1]Wang, DS; Hao, CH; Zheng, W; Peng, Q; Wang, TH; Liao, ZM; Yu, DP; Li, YD. Ultralong single-crystalline Ag_2S nanowires: promising candidates for photoswitches and room-temperature oxygen sensors. *Adv. Mater.,* 2008, 20, 2628–2632.

[423] Rout, CS; Govindaraj, A; Rao, CNR. High-sensitivity hydrocarbon sensors based on tungsten oxide nanowires. *J. Mater. Chem.,* 2006, 16, 3936-3941.

[424] Fu, XQ; Wang, C; Yu, HC; Wang, YG; Wang, TH. Fast humidity sensors based on CeO_2 nanowires. *Nanotechnol.,* 2007, 18, Art No. 145503.

[425] Wu, RJ; Sun, YL; Lin, CC: Chen HW, Chavali, M. Composite of TiO_2 nanowires and nafion as humidity sensor material. *Sensor Actuat. B,* 2006, 115, 198–204.

[426] Wang, G; Wang, Q; Lu, W; Li, JH. Photoelectrochemical study on charge transfer properties of TiO2-B nanowires with an application as humidity sensors. *J. Phys. Chem. B,* 2006, 110, 22029-22034.

[427] Rex, M; Hernandez, FE; Campiglia, AD. Pushing the limits of mercury sensors with gold nanorods. *Anal. Chem.,* 2006, 78, 445-451.

[428] Zhang, XP; Sun, BQ; Hodgkiss, JM; Friend, H. Tunable ultrafast optical switching via waveguided gold nanowires. *Adv. Mater.,* 2008, 20, 4455–4459.

[429] Li, ZL; Dharap, P; Nagarajaiah, S; Barrera, EV; Kim, JD. Carbon nanotubes film sensors. *Adv. Mater.,* 2004, 16, 640-643.

[430] Hasirci, V; Vrana, E; Zorlutuna, P; Ndreu, A; Yilgor, P; Basmanav, FB, Aydin, E. Nanobiomaterials: a review of the existing science and technology, and new approaches. *J. Niomater. Sci.-Polym. Ed.,* 2006, 17, 1241-1268.

[431] Goldberg, M; Langer, R; Jia, XQ. Nanostructured materials for applications in drug delivery and tissue engineering. *J. Biomater. Sci. – Polym. Ed.,* 2007, 18, 241-268.

[432] Salem, AK; Searson, PC; Leong, KW. Multifunctional nanorods for gene delivery. *Nature Mater.,* 2003, 2, 668-671.

[433] Huang, XH; El-Sayed, IH; Qian, W; El-Sayed, MA. Cancer cell imaging and photothermal therapy in the near-infrared region by using gold nanorods. *J. Am. Chem. Soc.,* 2006, 128, 2115-2120.

[434] Eghtedari, M; Oraevsky, A; Copland, JA; Kotov, NA; Conjusteau, A; Motamedi, M. High sensitivity of in vivo detection of gold nanorods using a laser optoacoustic imaging system. *Nano Lett.,* 2007, 7, 1914-1918.

[435] Durr, NJ; Larson, T; Smith, DK; Korgel, BA; Sokolov, K; Ben-Yakar, A. Two-photon luminescence imaging of cancer cells using molecularly targeted gold nanorods. *Nano Lett.,* 2007, 7, 941-945.

[436] Li, JL; Day, D; Gu, M. Ultra-low energy threshold for cancer photothermal therapy using transferrin-conjugated gold nanorods. *Adv. Mater.,* 2008, 20, 3866-3871.

[437] Niidome, T; Yamagata, M; Okamoto, Y; Akiyama, Y; Takahashi, H; Kawano, T; Katayama, Y; Niidome, Y. PEG-modified gold nanorods with a stealth character for in vivo applications. *J. Control. Release,* 2006, 114, 343–347.

[438] Wu, PC; Wang, WS; Huang, YT; Sheu, HS; Lo, YW; Tsai, TL; Shieh, DB; Yeh, CS. Porous iron oxide based nanorods developed as delivery nanocapsules. *Chem. Eur. J.*, 2007, 13, 3878 – 3885.

[439] Piao, Y; Kim, J; Na, HB; Kim, D; Baek, JS; Ko, MK; Lee, JH; Shokouhimehr, M; Hyeon, T. Wrap–bake–peel process for nanostructural transformation from -FeOOH nanorods to biocompatible iron oxide nanocapsules. Nature Mater., 2008, 7, 242-247.

[440] Giri, S; Trewyn, BG; Stellmaker, MP; Lin, VSY. Stimuli-responsive controlled-release delivery system based on mesoporous silica nanorods capped with magnetic nanoparticles. *Angew. Chem. Int. Ed.,* 2005, 44, 5038-5044.

[441] Pantarotto, D; Singh, R; McCarthy, D; Erhardt, M; Briand, JP; Prato, M; Kostarelos, K; Bianco, A. Functionalized carbon nanotubes for plasmid DNA gene delivery. *Angew. Chem. Int. Ed.,* 2004, 43, 5242-5246.

[442] Zhu, YH; Peng, AT; Carpenter, K; Maguire, JA; Hosmane, NS; Takagaki, M. Substituted carborane-appended water-soluble single-wall carbon nanotubes: New approach to boron neutron capture therapy drug delivery. *J. Am. Chem. Soc.,* 2005, 127, 9875-9880.

[443] Bianco, A; Kostarelos, K; Prato, M. Applications of carbon nanotubes in drug delivery. *Curr. Opinion Chem. Biology*, 2005, 9, 674–679.

[444] Reilly, RM. Carbon nanotubes: Potential benefits and risks of nanotechnology in nuclear medicine. *J. Nuclear Medicine,* 2007, 48, 1039-1042.

[445] Pastorin, G; Wu, W; Wieckowski, S; Briand, JP; Kostarelos, K; Prato, M; Bianco, A. Double functionalisation of carbon nanotubes for multimodal drug delivery. *Chem. Commun.,* 2006, (11), 1182-1184.

[446] Pantarotto, D; Singh, R; McCarthy, D; Erhardt, M; Briand, JP; Prato, M; Kostarelos, K; Bianco, A; Functionalized carbon nanotubes for plasmid DNA gene delivery. *Angew. Chem. Int. Ed.,* 2004, 43, 5242 –5246.

[447] Chen, CC; Liu, YC; Wu, CH; Yeh, CC; Su, MT; Wu, YC. Preparation of fluorescent silica nanotubes and their application in gene delivery. *Adv. Mater.,* 2005, 17, 404-407.

[448] Kumbar, SG; Nair, LS; Bhattacharyya, S; Laurencin, CT. Polymeric nanofibers as novel carriers for the delivery of therapeutic molecules. *J. Nanosci. Nanotechnol.,* 2006, 6, 2591-2607.

[449] Luong-Van, E; Grøndahl, L; Chua, KN; Leong, KW; Nurcombe, V; Cool, SM. Controlled release of heparin from poly(e-caprolactone) electrospun fibers. *Biomater.*, 2006, 27, 2042–2050.

[450] Cui, WG; Li, XH; Zhu, XL; Yu, G; Zhou, SB; Weng. J. Investigation of drug release and matrix degradation of electrospun poly(DL-lactide) fibers with paracetanol inoculation. *Biomacromol.,* 2006, 7, 1623-1629.

[451] Taepaiboon, P; Rungsardthong, U; Supaphol, P. Drug-loaded electrospun mats of poly(vinyl alcohol) fibres and their release characteristics of four model drugs. *Nanotechnol.,* 2006, 17, 2317-2329.

[452] Nikkola, L; Seppala, J; Harlin, A; Ndreu, A; Ashammakhi, N. Electrospun multifunctional diclofenac sodium releasing nanoscaffold. *J. Nanosci. Nanotechnol.,* 2006, 6, 3290-3295.

[453] Accardo, A; Tesauro, D; Mangiapia, G; Pedone, C; Morelli, G. Nanostructures by self-assembling peptide amphiphile as potential selective drug carriers. Biopolymers, 2007, 88, 115-121.

[454] Xie, JW; Wang, CH. Electrospun micro- and nanofibers for sustained delivery of paclitaxel to treat C6 glioma in vitro. *Pharm. Res.,* 2006, 23, 1817-1826.

[455] Song, M; Guo, DD; Pan, C; Jiang, H; Chen, C; Zhang, RY; Gu, ZZ; Wang, XM. The application of poly(N-isopropylacrylamide)-co-polystyrene nanofibers as an additive agent to facilitate the cellular uptake of an anticancer drug . Nanotechnol., 2008, 19, Art No. 165102.

[456] Zeng, J; Aigner, A; Czubayko, F; Kissel, T; Wendorff, JH; Greiner, A. Poly(vinyl alcohol) nanofibers by electrospinning as a protein delivery system and the retardation of enzyme release by additional polymer coatings. *Biomacromol.,* 2005, 6, 1484-1488.

[457] Valmikinathan, CM; Defroda, S; Yu, XJ. Polycaprolactone and bovine serum albumin based nanofibers for controlled release of nerve growth factor. , *Biomacromolecules,* 2009, *10,* 1084–1089

[458] Kim, TG; Lee, DS; Park, TG. Controlled protein release from electrospun biodegradable fiber mesh composed of poly(ε-caprolactone) and poly(ethylene oxide). *Int. J. Pharm.,* 2007, 338, 276-283.

[459] Yang, ZM; Xu, KM; Wang, L; Gu, HW; Wei, H; Zhang, MJ; Xu, B. Self-assembly of small molecules affords multifunctional supramolecular hydrogels for topically treating simulated uranium wounds. *Chem. Commun.,* 2005, (35), 4414-4416.

[460] Duan, YY; Jia, J; Wang, SH; Yan, W; Jin, L; Wang, ZY. Preparation of antimicrobial poly(epsilon-caprolactone) electrospun nanofibers containing silver-loaded zirconium phosphate nanoparticles. *J. Appl. Polym. Sci.*, 2007, 106, 1208-1214.

[461] Ignatova, M; Manolova, N; Rashkov, I. Electrospinning of poly(vinyl pyrrolidone)–iodine complex and poly(ethylene oxide)/poly(vinyl pyrrolidone)–iodine complex – a prospective route to antimicrobial wound dressing materials. *Euro. Polym. J.,* 2007, 43, 1609-1623.

[462] Hong, KH. Preparation and properties of electrospun poly (vinyl alcohol)/silver fiber web as wound dressings. *Polym. Eng. Sci.,* 2007, 47, 43-49.

[463] Hong, KH; Park, JL; Sul, IH; Youk, JH; Kang, TJ. Preparation of antimicrobial poly(vinyl alcohol) nanofibers containing silver nanoparticles. *J. Polym. Sci. B- Polym. Phys.*, 2006, 44, 2468-2474.

[464] Jo, E; Lee, S; Kim, KT; Won, YS; Kim, HS; Cho, EC; Jeong, U. Core-sheath nanofibers containing colloidal arrays in the core for programmable multi-agent delivery. *Adv. Mater.,* 2009, 21, 968–972

[465] Kim, HW; Song, JW; Kim, HE. Nanofiber generation of gelatin-hydroxyapatite biomimetics for guided tissue regeneration. *Adv. Funct. Mater.,* 2005, 15, 1988-1994.

[466] Hosseinkhani, H; Hosseinkhani, M; Khademhosseini, A; Kobayashi, H. Bone regeneration through controlled release of bone morphogenetic protein-2 from 3-D tissue engineered nano-scaffold. *J. Control. Release,* 2007, 117, 380-386.

[467] Goldberg, M; Langer, R; Jia, X Q. Nanostructured materials for applications in drug delivery and tissue engineering. *J. Biomater. Sci., Polym. Ed.,* 2007, 18, 241-268.

[468] Prabhakaran, MP; Venugopal, J; Chan, CK; Ramakrishna, S. Surface modified electrospun nanofibrous scaffolds for nerve tissue engineering. Nanotechnol., 2008, 19, Art No. 455102.

[469] Fu, YC; Nie, H; Ho, ML; Wang, CK; Wang, CH. Optimized bone regeneration based on sustained release from three-dimensional fibrous PLGA/HAp composite scaffolds loaded with BMP-2. *Biotechnol. Bioeng.* 2008, 99, 996-1006.

[470] Zhang, YZ; Venugopal, JR; El-Turki, A; Ramakrishna, S; Su, B; Lim, CT. Electrospun biomimetic nanocomposite nanofibers of hydroxyapatite/chitosan for bone tissue engineering. *Biomater.,* 2008, 29, 4314–4322.

[471] Yu, MF; Lourie, O; Dyer, MJ; Moloni, K; Kelly, TF; Ruoff, RS. Strength and breaking mechanism of multiwalled carbon nanotubes under tensile load. *Science*, 2000, 287, 637 - 640.

[472] Mordkovich, VZ. Carbon nanofibers: a new ultrahigh-strength material for chemical technology. *Theoretical Foundations Chem. Eng.,* 2003, 37, 429-438.

[473] Ye, HH; Lam, H; Titchenal, N; Gogotsi, Y; Ko, F. Reinforcement and rupture behavior of carbon nanotubes–polymer. *Appl. Phys. Lett.,* 2004, 6, 1775-1777.

[474] Bokobza, L; Chauvin, LP. Reinforcement of natural rubber: use of in situ generated silicas and nanofibres of sepiolite. *Polym.,* 2005, 46, 4144-4151.

[475] Blake, R; Coleman, JN; Byrne, MT; McCarthy, JE; Perova, TS; Blau, WJ; Fonseca, A; Nagy, JB; Gun'ko, YK. Reinforcement of poly(vinyl chloride) and polystyrene using chlorinated polypropylene grafted carbon nanotubes. *J. Mater. Chem.*, 2006, 16, 4206-4213.

[476] Cha, SI; Kim, KT; Arshad, TSN; Mo, CB; Hong, SH; Extraordinary strengthening effect of carbon nanotubes in metal-matrix

nanocomposites processed by molecular-level mixing. *Adv. Mater.,* 2005, 17, 1377-1381.

[477] Ayutsede, J; Gandhi, M; Sukigara, S; Ye, HH; Hsu, CM; Gogotsi, Y; Ko, F. Carbon nanotube reinforced Bombyx mori silk nanofibers by the electrospinning process. *Biomacromol.,* 2006, 7, 208-214.

[478] Liu, LQ; Tasis, D; Prato, M; Wagner, HD. Tensile mechanics of electrospun multiwalled nanotube/poly(methyl methacrylate) nanofibers. *Adv. Mater.,* 2007, *19*, 1228–1233.

[479] Zhu, J; Peng, HQ; Rodriguez-Macias, F; Margrave, JL; Khabashesku, VN; Imam, AM; Lozano, K; Barrera, EV. Reinforcing epoxy polymer composites through covalent integration of functionalized nanotubes. *Adv. Funct. Mater.,* 2004, 14, 643-648.

[480] Spinks, GM; Mottaghitalab, V; Bahrami-Samani, M; Whitten, PG; Wallace, GG. Carbon-nanotube-reinforced polyaniline fibers for high-strength artificial muscles. *Adv. Mater.,* 2006, *18*, 637–640.

[481] Chena, MY; Baib, Z; Tan, SC; Unroe, MR. Friction and wear scar analysis of carbon nanofiber-reinforced polymeric composite coatings on alumina/aluminum composite. *Wear*, 2002, 252, 624–634.

[482] Kelarakis, A; Yoon, K; Somani, R; Sics, I; Chen, XM; Hsiao, BS; Chu, B. Relationship between structure and dynamic mechanical properties of a carbon nanofiber reinforced elastomeric nanocomposite. *Polym.*, 2006, 47, 6797-6807.

[483] Cho, MH; Bahadur, S. A study of the thermal, dynamic mechanical, and tribological properties of polyphenylene sulfide composites reinforced with carbon nanofibers. *Tribology Lett.,* 2007, 25, 237-245.

[484] Chen, Q; Tan, JGH; Shen, SC; Liu, YC; Ng, WK; Zeng, XT. Effect of boehmite nanorods on the properties of glycidoxypropyltrimethoxysilane (GPTS) hybrid coatings *J. Sol-Gel. Sci. Technol.,* 2007, 44, 125-131.

[485] Chen, Q; Udomsangpetch, C; Shen, SC; Liu, YC; Chen, Z; Zeng, XT. The effect of AlOOH boehmite nanorods on mechanical property of hybrid composite coatings. , *Thin Solid Films,* 2009, 517, 4871–4874.

[486] Kobayashi, S; Kawai, W. Development of carbon nano.ber reinforced hydroxyapatite with enhanced mechanical properties. *Composites: Part A,* 2007, 38, 114–123.

[487] Balani, K; Anderson, R; Laha, T; Andara, M; Tercero, J; Crumpler, E; Agarwal, A; Plasma-sprayed carbon nanotube reinforced hydroxyapatite coatings and their interaction with human osteoblasts in vitro. *Biomater.,* 2007, 28, 618-624.

[488] Ross, DK. Hydrogen storage: The major technological barrier to the development. of hydrogen fuel cell cars. *Vcauum,* 2006, 80, 1084-1089.

[489] Marella, M; Tomaselli, M. Synthesis of carbon nanofibers and measurements of hydrogen storage. *Carbon*, 2006, 8, 1404-1413.

[490] Simonyan, VV; Johnson, JK. Hydrogen storage ion carbon nanotubes and graphite nanofibers. *J. Alloys Compounds,* 2002, 330-332, 659-665.

[491] Bououdina, M; Grant, D; Walker, G. Review on hydrogen absorbing materials - structure, microstructure, and thermodynamic properties. *Int. J. Hydrogen Energy,* 2006, 31, 177-182.

[492] Mishra, A; Banerjee, S; Mohapatra, SK; Graeve, OA; Misra, M. Synthesis of carbon nanotube-TiO2 nanotubular material for reversible hydrogen storage. *Nanotechnol.,* 2008, 19, Art No. 445607.

[493] Chambers, A; Park, C; Terry, R; Baker, K; Rodriguez, NM. Hydrogen Storage in Graphite Nanofibers. *J. Phys. Chem. B, 1998, 102,* 4254-4256.

[494] Browning, DJ; Gerrard, ML; Lakeman, JB; Mellor, IM; Mortimer, RJ; Turpin, MC. Studies into the storage of hydrogen in carbon nanofibers: Proposal of a possible reaction mechanism. *Nano Lett.,* 2002, 2, 201-205.

[495] Darkrima, FL; Malbrunot, P; Tartaglia, GP. Review of hydrogen storage by adsorption in carbon nanotubes. *Int. J. Hydrogen Energy,* 2002, 27, 193-202.

[496] Zhou, YP; Feng, K; Sun, Y; Zhou, L. A brief review on the study of hydrogen storage in terms of carbon nanotubes. *Prog. Chem.,* 2003, 15, 345-350.

[497] Chen, J; Wu, F. Review of hydrogen storage in inorganic fullerene-like nanotubes. *Appl. Phys. A,* 2004, 78, 989–994.

[498] [1]Zuttel, A; Sudan, P; Mauron, P; Kiyobayashi, T; Emmenegger, T; Schlapbach, L. Hydrogen storage in carbon nanostructures. *Int. J. Hydrogen Storage*, 2002, 27, 203–212.

[499] Sankaran, M; Viswanathan, B. The role of heteroatoms in carbon nanotubes for hydrogen storage. *Carbon*, 2006, 44, 2816-2821.

[500] Mu, SC; Tang, HL; Qian, SH; Pan, M; Yuan, RZ. Hydrogen storage in carbon nanotubes modified by microwave plasma etching and Pd decoration. *Carbon*, 2006, 44, 762-767.

[501] Lee, JW; Kim, HS; Lee, JY; Kang, JK. Hydrogen storage and desorption properties of Ni-dispersed carbon nanotubes. *Appl. Phys. Lett.,* 2006, 88, Art No. 143126.

[502] Zhang, BY; Liang, QM,; Song, C; Xia, YY; Zhao, MW; Liu, XD; Zhang, HY. Hydrogen storage in benzene moiety decorated single-walled carbon nanotubes. *Chin. Phys. Lett.,* 2006, 23, 1536-1539.

[503] Jung, JH; Rim, JA; Lee, SJ; Cho, SJ; Kim, SY; Kang, JK; Kim, YM; Kim, YJ. Pd-doped double-walled silica nanotubes as hydrogen storage material at room temperature. *J. Phys. Chem. C*, 2007, 111, 2679-2682.

[504] Ren, JW; Liao, SJ; Liu, JM. Hydrogen storage of multiwalled carbon nanotubes coated with Pd-Ni nanoparticles under moderate conditions. *Chin. Sci. Bull.,* 2006, 51, 2959-2963.

[505] Chen, CH; Huang, CC. Hydrogen storage by KOH-modified multi-walled carbon nanotubes. *Int. J. Hydrogen Energy*, 2007, 32, 237-246.

[506] Cho, JH; Park, CR. Hydrogen storage on Li-doped single-walled carbon nanotubes: Computer simulation using the density functional theory. *Catal. Today*, 2007, 120, 407-412.

[507] Sankaran, M; Viswanathan, B. Hydrogen storage in boron substituted carbon nanotubes. *Carbon*, 2007, 45, 1628-1635.

[508] Felderhoff, M; Weidenthaler, C; Helmolt RV; Eberle, U. Hydrogen storage: the remaining scientific and technological challenges. *Phys. Chem. Chem.*, Phys., 2007, 9, 2643–2653.

[509] Cheng, FY; Chen, J. Storage of hydrogen and lithium in inorganic nanotubes and nanowires. *J. Mater. Res.,* 2006, 21, 2744-2757.

[510] Meng, S; Kaxiras, E; Zhang, ZY. Metal-diboride nanotubes as high-capacity hydrogen storage media. *Nano Lett.,* 2007, 7, 663-667.

[511] Mpourmpakis, G; Froudakis, GE. Why boron nitride nanotubes are preferable to carbon nanotubes for hydrogen storage? An ab initio theoretical study. *Catal. Today,* 2007, 120, 341-345.

[512] Kim, SY; Kim, HS; Augustine, S; Kang, JK. Nanopores in carbon nitride nanotubes: Reversible hydrogen storage sites. *Appl. Phys. Lett.*, 2006, 89, Art. No. 253119.

[513] Jhi, SH. Activated boron nitride nanotubes: A potential material for room-temperature hydrogen storage. *Phys. Rev.*, 2006, 74, Art. No. 155424.

[514] Mpourmpakis, G; Froudakis, GE; Lithoxoos, GP; Samios, J. SiC nanotubes: A novel material for hydrogen storage. *Nano Lett.*, 2006, 6, 1581-1583.

[515] Ji, LW; Zhang, XW. Fabrication of porous carbon nanofibers and their application as anode materials for rechargeable lithium-ion batteries. Nanotechnol. 2009, 20, Art No 155705.

[516] Zhang, J; Hu, YS; Tessonnier, JP; Weinberg, G; Maier, J; Schlögl, R; Su, DS. CNFs@CNTs: superior carbon for electrochemical energy storage. *Adv. Mater.,* 2008, 20, 1450–1455.

[517] Fischer, T; Hampp, NA. Encapsulation of purple membrane patches into polymeric nanofibers by electrospinning. *IEEE Trans. Nanobiosci.* 2004, 3, 118-120.

[518] Shin, C; Chase, GG. Separation of water-in-oil emulsions using glass fiber media augmented with polymer nanofibers. *J. Dispers. Sci. Tech.,* 2006, 27, 517-522.

[519] Zhang, LF; Hsieh, YL. Nanoporous ultrahigh specific surface polyacrylonitrile fibres. *Nanotechnol.* 2006, 17, 4416-4423.

[520] Qin, XH; Wang, SY. Filtration properties of electrospinning nanofibers. *J. Appl. Polym. Sci.*, 2006, 102, 1285-1290.

[521] Gopal, R; Kaur, S; Ma, ZW; Chan, C; Ramakrishna, S; Matsuura, T. Electrospun nanofibrous filtration membrane. *J. Membane Sci.*, 2006, 281, 581-586.

[522] Yang, Q; Wu, J; Li, JJ; Hu, MX; Xu, ZK. Nanofibrous sugar sticks electrospun from glycopolymers for protein separation via molecular recognition. *Macromol. Rapid Commun.,* 2006, 27, 1942–1948.

[523] Ma, ZW; Kotaki, M; Ramakrishna, S. Electrospun cellulose nanofiber as affinity membrane. *J. Membrane Sci.,* 2005, 265, 115–123.

[524] Ke, XB; Zhu, HY; Gao, XP; Liu, JW; Zheng, ZF. High-performance ceramic membranes with a separation layer of metal oxide nanofibers. *Adv. Mater.,* 2007, *19*, 785–790.

[525] Peng, XS; Jin, J; Ichinose, I. Mesoporous separation membranes of polymer-coated copper hydroxide nanostrands. *Adv. Funct. Mater.,* 2007, 17, 1849–1855.

[526] Podgorski, A; Balazy, A; Gradon, L. Application of nanofibers to improve the filtration efficiency of the most penetrating aerosol particles in fibrous filters. *Chem. Eng. Sci.,* 2006, 61, 6804-6815.

[527] Viswannathan, G; Kane, DB; Lipowicz, PJ. High efficiency fine particulate filtration using carbon nanotube coatings. *Adv. Mater.,* 2004, 16, 2045-2049.

[528] Yang, DJ; Zheng, ZF; Zhu, HY; Liu, HW; Gao, XP. Titanate nanofibers as intelligent absorbents for the removal of radioactive ions from water. *Adv. Mater.,* 2008, 20, 2777–2781.

[529] Shen, SC; Hidajat, K; Yu, LE; Kawi, S. A Simple Hydrothermal Synthesis of Nanostructured and Nanorod Zn-Al complex oxides as Novel NanoCatalysts, *Adv. Mater.*, 2004, 16, 541-545.

[530] Chuangchote, S; Jitputti, J; Sagawa, T; Yoshikawa, S. Photocatalytic activity for hydrogen evolution of electrospun TiO_2 nanofibers. *Appl. Mater. Interface.,* 2009, 1, 1140-1143.

[531] Kwak, G; Yong, K. Adsorption and reaction of ethanol on ZnO nanowires. *J. Phys. Chem. C,* 2008, *112,* 3036-3041.

[532] Zhou, KB; Wang, X; Sun, XM; Peng, Q; Li, YD. Enhanced catalytic activity of ceria nanorods from well-defined reactive crystal planes. *J. Catal.*, 2005, 229, 206-212.

[533] Formo, E; Lee, E; Campbell, D; Xia, YN. Functionalization of electrospun TiO_2 nanofibers with Pt nanoparticles and nanowires for catalytic applications. Nano Lett. 2008, 8, 668-672.

[534] Xie, XW; Li, Y; Liu, ZQ; Haruta, M; Shen, WJ. Low-temperature oxidation of CO catalysed by Co_3O_4 nanorods. *Nature*, 2009, 458, 746-749.

[535] Lee, JA; Krogman, KC; Ma, ML; Hill, RM; Hammond, PT; Rutledge, GC. Highly reactive multilayer-assembled TiO_2 coating on electrospun polymer nanofibers. *Adv. Mater.,* 2009, 21, 1252–1256.

[536] Lin, DD; Wu, H; Zhang, R; Pan, W. Enhanced photocatalysis of electrospun Ag-ZnO heterostructured nanofibers. *Chem. Mater.* ,2009, 21, 3479–3484.

[537] Li, H; Lin, H; Xie, SH; Dai, WL; Qiao, MH; Lu, YF; Li, HX. Ordered Mesoporous Ni Nanowires with Enhanced Hydrogenation Activity Prepared by Electroless Plating on Functionalized SBA-15. *Chem. Mater.,* 2008, *20,* 3936–3943.

[538] Bezemer, GL; Radstake, PB; Koot, V; van Dillen, AJ; Geus, JW; de Jong, KP. Preparation of Fischer-Tropsch cobalt catalysts supported on carbon nanofibers and silica using homogeneous deposition-precipitation. *J. Catal.*, 2006, 237, 291-302.

[539] Winter. F; Koot, V; van Dillen, AJ; Geus, JW; de Jong, KP. Hydrotalcites supported on carbon nanofibers as solid base catalysts for the synthesis of MlBK. *J. Catal.*, 2005, 236, 91-100.

[540] Guczi, L; Stefler, G; Geszti, O; Koppany, Z; Konya, Z; Molnar, E; Urban, M; Kiricsi, I. CO hydrogenation over cobalt and iron catalysts supported over multiwall carbon nanotubes: Effect of preparation. *J. Catal.*, 2006, 244, 24-32.

[541] Delgado, JJ; Su, DS; Rebmann, G; Keller, N; Gajovic, A; Schlogl, R. Immobilized carbon nanofibers as industrial catalyst for ODH reactions. *J. Catal.*, 2006, 244, 126-129.

[542] Zhong, ZY; Ho, J; Teo, J; Shen, SC; Gedanken, A. Synthesis of Porous r-Fe_2O_3 nanorods and deposition of very small gold particles in the pores for catalytic oxidation of CO. *Chem. Mater.* 2007, *19,* 4776-4782.

[543] Wang, HJ; Zhou, AL; Peng, F; Yu, H; Chen, LF. Adsorption characteristic of acidified carbon nanotubes for heavy metal Pb(II) in aqueous solution. *Mater. Sci. Eng. A- Struct. Mater. Properties Microstruct. Process.,* 2007, 466, 210-206.

[544] Diaz, E; Ordonez, S; Vega, A. Adsorption of volatile organic compounds onto carbon nanotubes, carbon nanofibers, and high-surface-area graphites. *J. Coloid Interface Sci.,* 2007, 305, 7-16.

[545] Zhu, K; Neale, NR; Miedaner, A; Frank, AJ. Enhanced charge-collection efficiencies and light scattering in dye-sensitized solar cells using oriented TiO_2 nanotubes arrays. *Nano Lett.,* 2007, 7, 69-74.

[546] Lee, W; Lee, J; Lee, H; Yi, W; Han, SH. Enhanced charge-collection efficiency of In_2S_3/In_2O_3 photoelectrochemical cells in the presence of single-walled carbon nanotubes. *Appl. Phys. Lett.,* 2007, 91, Art. No. 043515.

[547] Dong, H; Wang, D; Sun, G; Hinestroza, JP. Assembly of metal nanoparticles on electrospun nylon 6 nanofibers by control of interfacial hydrogen-bonding interactions. *Chem. Mater.,* 2008, 20, 6627–6632.

[548] Lin, WS; Huang, YW; Zhou, XD; Ma, YF. In vitro toxicity of silica nanoparticles in human lung cancer cells. *Toxic. Appl. Pharm.,* 2006, 217, 252-259.

[549] Dobrovolskaia, MA; McNeil, SE. Immunological properties of engineered nanomaterials. *Nature Nanotechnol.,* 2007, 2, 469-478.

[550] Jeng, HA; Swanson, J. Toxicity of metal oxide nanoparticles in mammalian cells. *J. Environ. Sci. Health Part A- Toxic/Hazardous Subatance Environ. Eng.,* 2006, 41, 2699-2711.

[551] Jin, YH; Kannan, S; Wu, M; Zhao, JXJ. Toxicity of luminescent silica nanoparticles to living cells. *Chem. Res. Toxic.,* 2007, 20, 1126-1133.

[552] Zhang, YB; Chen, W; Zhang, J; Liu, J; Chen, GP; Pope, C. In vitro and in vivo toxicity of CdTe nanoparticles. *J. Nanosci. Nanotechnol.,* 2007, 7, 497-503.

[553] Sayes, CM; Reed, KL; Warheit, DB. Assessing toxicity of fine and nanoparticles: Comparing in vitro measurements to in vivo pulmonary toxicity profiles. *Toxic. Sci.,* 2007, 97, 163-180.

[554] Maynard, AD; Aitken, RJ; Butz, T; Colvin, V; Donaldson, K; Oberdorster, G; Philbert, MA; Ryan, J; Seaton, A; Stone, V; Tinkle, SS;

Tran, L; Walker, NJ; Warheit, DB. Safe handling of nanotechnology. *Nature*, 2006, 444, 267–269.

[555] Boczkowski, J; Lanone, S. Potential uses of carbon nanotubes in the medical field: how worried should patients be? *Nanomedicine*, 2007, 2, 407-410.

[556] Seo, JW; Magrez, A; Milas, M; Lee, K; Lukovac, V; Forro, L. Catalytically grown carbon nanotubes: from synthesis to toxicity. *J. Phys. D- Appl. Phys.,* 2007, 40, R109-R120.

[557] Warheit, DB; Laurence, BR; Reed, KL; Roach, DH; Reynolds, GAM; Webb, TR. Comparative pulmonary toxicity assessment of single-wall carbon nanotubes in rats. *Toxic. Sci.,* 2004, 77, 117–125.

[558] Magrez, A; Kasas, S; Salicio, V; Pasquier, N; Seo, JW; Celio, M; Catsicas, S; Schwaller, B; Forro, L. Cellular toxicity of carbon-based nanomaterials. *Nano Lett.,* 2006, 6, 1121-1125.

[559] Mullera, J; Huauxa, F; Moreaub, N; Missona, P; Heiliera, JF; Delosc, M; Arrasa, M; Fonsecab, A; Nagyb, JB; Lison, D. Respiratory toxicity of multi-wall carbon nanotubes. *Toxico. Appl. Pharm.*, 2005, 207, 221–231.

[560] Smith, CJ; Shaw, BJ; Handy, RD. Toxicity of single walled carbon nanotubes to rainbow trout, (Oncorhynchus mykiss): Respiratory toxicity, organ pathologies, and other physiological effects. *Aquatic Toxic.*, 2007, 82. 94–109.

[561] Davoren, M; Herzog, E; Casey, A; Cottineau, B; Chambers, G; Byrne, HJ; Lyng, FM. In vitro toxicity evaluation of single walled carbon nanotubes on human A549 lung cells. *Toxic. in Vitro*, 2007, 21, 438–448.

[562] Pulskamp, K; Diabate, S; Krug, HF. Carbon nanotubes show no sign of acute toxicity but induce intracellular reactive oxygen species in dependence on contaminants. *Toxic. Lett.*, 2007, 168, 58-74.

[563] Brown, DM; Kinloch, IA; Bangert, U; Windle, AH; Walter, DM; Walker, GS; Scotchford, CA; Donaldson, K; Stone, V. An in vitro study of the potential of carbon nanotubes and nanofbres to induce inflammatory mediators and frustrated phagocytosis. *Carbon*, 2007, 45, 1743–1756.

[564] Donaldson, K; Aitken, R; Tran, L; Stone, V; Duffin, R; Forrest, G; Alexander, A. Carbon nanotubes: a review of their properties in relation to pulmonary toxicology and workplace safety. *Toxic. Sci.*, 2006, 92, 5-22.

INDEX

A

absorbents, 169
acetate, 54
acetylene, 26, 155
acid, 27, 46, 50, 72, 88, 95, 97, 100, 107, 118
acidic, 62
acidification, 143
acidity, 150
actuation, 108
actuators, 157
acute, 118, 119, 171
administration, 100, 118
adsorption, 111, 112, 166
aerosol, 115, 168
aerospace, 109
AFM, 79
agent, 47, 48, 99, 102, 107, 163, 164
agents, 55, 99
aggregates, 71, 72
aggregation, 119, 120
aid, 104, 123
air, 16, 37, 39, 63, 91, 115
airborne particles, 117
airways, 118, 119
alcohol, 49, 103, 151, 162, 163, 164
alkaline, 26, 62, 134
alpha, 141, 143, 150, 160
alternative, 97, 100, 113
aluminium, 134, 157
aluminosilicate, 150
aluminum, xvi, 23, 27, 45, 52, 59, 62, 65, 76, 134, 136, 144, 150, 159, 165
aluminum oxide, xvi, 23, 134, 159
alveolitis, 119
ambient pressure, 49
American Association for the Advancement of Science, 9, 19, 33, 71, 87
amine, 50, 95, 140
amino, 97, 102
amino acid, 102
amino acids, 102
amino groups, 97
ammonia, 61, 149
ammonium, 45, 143
amorphous, 9, 21, 25, 39, 41, 55, 59, 60, 62, 63, 65, 146
amorphous carbon, 146
amorphous precipitate, 59, 63
anatase, 70, 135, 143
angiogenic, 101
animals, 120
annealing, 152
Annealing, 78
anode, 113, 168
anodes, 113
antibacterial, 116

antibody, 89
anticancer, 98, 99, 100, 163
anticancer drug, 98, 99, 100, 163
antigen, 89
antimony, 128, 158
aqueous solution, 27, 42, 43, 45, 46, 55, 61, 73, 142, 170
argon, 13, 21
aspect ratio, xv, 14, 15, 34, 41, 43, 46, 48, 61, 72, 109, 145
assessment, 120, 171
atmosphere, 14, 20, 32, 33, 52, 78
atomic force, 76
atomic force microscopy, 76
atoms, 13, 59, 60, 63, 113
attachment, 154
Au nanoparticles, 27, 34, 35
automotive application, 111
automotive applications, 111
availability, 9

B

back, 79
bandgap, 77
barium, 54
barrier, 84, 166
batteries, 113, 168
battery, 35, 137, 159, 160
behavior, 99, 139, 164
bending, 62, 84, 86, 109
benefits, 117, 162
benzene, 113, 146, 167
bias, 82, 83, 84
binding, 27, 96, 113
bioavailability, 105
biocompatibility, 110
biocompatible, 97, 103, 162
biodegradable, 100, 101, 102, 103, 163
biodegradable materials, 103
bioengineering, 116
biological activity, 104
biological systems, 115
biomacromolecules, 105
biomaterials, 47, 89, 106, 109
biomedical applications, 95, 97
biomimetic, 164
biomolecular, 46
biomolecules, 89, 114, 123
Biopolymers, 163
Biosensor, 158
biosensors, 89, 158, 159
bismuth, 7, 46, 126, 140, 144, 153
blocks, 28, 76, 80
blood, 81, 96
blood plasma, 81
body fluid, 47
boiling, 48
bonding, 45, 132, 170
bonds, 21
bone growth, 103
Boron, 25, 89, 113
boron neutron capture therapy, 162
boron nitride nanotubes, 167
bottom-up, 1, 67
bovine, 163
brain, 100
brain tumor, 100
branching, 46, 62, 65
buffer, 88, 90, 101
building blocks, 28, 76
buildings, 106
bulk materials, 75, 95, 105
burnout, 57

C

cables, 52
cancer, 96, 98, 101, 161, 170
cancer cells, 96, 101, 161, 170
candidates, 25, 93, 97, 110, 160
capillary, 27, 70, 151
caprolactone, 100, 101, 162, 163
carbide, 32, 113, 137
carbide nanorods, 137
carbon, iii, 26, 31, 32, 33, 34, 35, 37, 39, 40, 49, 72, 73, 76, 77, 79, 80, 89, 91, 94, 97, 98, 99, 105, 106, 107, 108, 109, 111, 112, 115, 118, 119, 120, 133, 135, 136, 137, 138, 139, 141, 146, 152, 154, 155,

156, 159, 161, 162, 164, 165, 166, 167, 168, 169, 170, 171, 172
carbon materials, 109
carbon nanotubes, 26, 31, 32, 33, 34, 35, 39, 40, 49, 76, 77, 79, 80, 89, 91, 95, 97, 98, 99, 105, 106, 107, 110, 111, 112, 113, 115, 118, 119, 120, 135, 136, 137, 146, 152, 155, 156, 159, 161, 162, 164, 165, 166, 167, 170, 171
carcinogenic, 120
carrier, 21, 97, 98, 157
catalysis, xv, 116
catalyst, 2, 3, 4, 6, 7, 8, 9, 10, 11, 13, 19, 22, 50, 66, 73, 116, 123, 130, 155, 170
catalytic activity, 169
category d, 123
cell, 60, 95, 97, 102, 110, 118, 119, 161, 166
cell death, 118
cell growth, 110
cell membranes, 97
Cellular response, 119
cellulose, 114, 115, 168
cement, 106
ceramic, 22, 51, 114, 152, 168
ceramics, xv, 54, 61, 70, 110, 148
cerium, 141, 154
CH4, 91, 139
channels, 23, 25, 26, 27, 31, 41, 79, 132
chemical approach, 146
chemical bonds, 109
chemical etching, 41
chemical properties, 103, 123
chemical vapor deposition, 12, 19, 36, 126, 128, 132, 155
chemicals, 41, 88
chemiluminescence, 34
chitosan, 89, 104, 151, 158, 164
chloride, 47, 72, 106, 165
chloroform, 26
chromium, 3, 6
cladding, 80
classes, 105
clusters, 14, 23, 31, 130, 135
CNTs, 168
CO2, 33
coatings, 108, 163, 165, 166, 168
cobalt, 3, 6, 47, 72, 145, 154, 169
collagen, 119
colloids, 103
complementary DNA, 88
components, 9, 34, 36, 88, 94, 96, 109
composites, 33, 106, 107, 109, 137, 165
composition, 2, 75, 76, 123
compound semiconductors, 51
compounds, 9, 43
computation, 85
Computer simulation, 167
concentration, 45, 88, 90, 92
condensation, 10, 60
conducting polymers, 80, 157
conduction, 156
conductive, 152, 154
conductivity, 27, 77, 78, 80, 87, 94
configuration, 30, 156
confinement, 111, 142
Congress, x
conjugation, 98
construction, 106
consumption, 36
contaminants, 171
contamination, 12, 120
continuity, 41
control, 36, 45, 50, 55, 77, 80, 84, 87, 88, 90, 100, 103, 105, 117, 121, 133, 143, 147, 170
conversion, 46, 55, 65, 66, 68, 141, 142, 150
copolymer, 108, 138
copper, 3, 7, 9, 46, 48, 115, 140, 145, 146, 168
core-shell, 103, 151, 160
cost-effective, 42
costs, 124
coupling, 83
covalent, 107, 165
covalent bond, 107
crack, 109
cross-linking, 72, 107
cross-linking reaction, 72

cross-sectional, 79
crystal growth, 15, 19, 26, 42, 48, 55, 59, 67, 125
crystal structure, 18, 46
crystal structures, 46
crystalline, 9, 14, 26, 42, 47, 49, 60, 72, 126, 127, 132, 133, 134, 136, 139, 143, 144, 145, 147, 160
crystalline solids, 60
crystallinity, 45, 46, 75, 108, 110, 123
crystallites, 148
crystallization, 2, 42, 47, 59, 60, 62, 124, 144
crystals, 55
CTAB, 46, 52
cubic boron nitride, 149
curing, 107
current ratio, 84
CVD, 19, 20, 21, 22, 25, 36, 41, 123
cycles, 28, 37, 113
cyclic voltammetry, 89
cytokine, 119
cytoplasm, 97
cytotoxicity, 96, 105, 117

D

dairy, 116
death, 118
decomposition, 13, 19, 26, 33, 42, 43, 73, 155
defects, 104, 112
definition, xvi
deformation, 84
degradation, 50, 162
dehydration, 60
delivery, 95, 96, 97, 98, 99, 101, 102, 103, 105, 118, 120, 123, 161, 162, 163, 164
density, 3, 6, 25, 76, 79, 112, 133, 167
density functional theory, 167
Department of Energy (DOE), 112
deposition, 1, 4, 5, 10, 12, 19, 20, 21, 22, 23, 25, 28, 32, 34, 36, 73, 76, 79, 126, 127, 128, 132, 133, 134, 135, 136, 138, 139, 155, 169, 170
derivatives, 159
desorption, 167
destruction, 103
detection, 87, 89, 90, 93, 158, 159, 160, 161
dialysis, 115
Diamond, 131, 152
dielectric constant, 77
dielectrics, 77, 156
diffraction, 60
diffusion, 11, 23, 129, 132
digestion, 41
dimensionality, xv
diodes, 9, 83
dipole, 72
disorder, 150
dispersion, 48, 106, 107, 113
displacement, 131
distilled water, 27, 37
distribution, 96, 98, 110
DNA, 35, 87, 88, 89, 90, 95, 97, 98, 105, 137, 158, 162
dopant, 72
doped, 144, 149, 167
doping, 113, 153
dressing material, 163
dressings, 102, 163
drug carriers, 97, 100, 163
drug delivery, 96, 97, 98, 99, 103, 105, 118, 120, 123, 161, 162, 164
drug release, 100, 162
drug-resistant, 101
drugs, 97, 101, 103, 105, 162
drying, 27, 31
dyes, 115

E

earth, 26, 139
elasticity, 126
electric conductivity, 27
electric current, 84
electric potential, 85
electrical conductivity, 80, 87
electrical properties, 126, 157

electrical resistance, 90
electroactivity, 108
electrodeposition, 25, 31, 133, 135
electrodes, 35, 137, 159
electroless deposition, 28, 134, 135, 139
electrolyte, 80
electron, 6, 49, 77, 79, 94
electron microscopy, 6, 49, 79
electrospinning, xvi, 54, 67, 69, 70, 99, 101, 102, 104, 107, 114, 138, 149, 151, 152, 153, 163, 165, 168
electrostatic force, 67
emission, 6, 14, 82, 83, 131, 138
emulsions, 168
encapsulated, 100, 104, 105, 113
encapsulation, 100, 151
energy, 7, 75, 77, 84, 111, 113, 157, 161, 168
environment, 13, 16, 20, 21, 53, 59, 62, 81, 84, 103, 110
environmental impact, 41, 117, 124
enzymatic, 89
enzymes, 102
epoxy, 107, 165
epoxy polymer, 107, 165
equilibrium, 4
etching, 26, 31, 41, 112, 138, 167
ethanol, 33, 48, 52, 70, 91, 93, 146, 149, 159, 160, 169
ethylene, 45, 101, 108, 141, 144, 151, 152, 163
ethylene glycol, 141
ethylene oxide, 45, 101, 144, 151, 152, 163
ethylenediamine, 141
Euro, 140, 145, 151, 163
evaporation, 5, 13, 19, 20, 21, 22, 71, 126, 127, 131, 153, 157
evolution, 169
excitation, 94
excretion, 96
exposure, 88, 117, 119, 120
extraction, 70, 147
extrusion, 136

F

fabricate, 41, 52, 89, 113
fabrication, xv, 1, 5, 8, 9, 19, 22, 23, 25, 26, 27, 30, 32, 36, 41, 51, 59, 66, 67, 73, 75, 76, 77, 80, 99, 123, 130, 137, 138, 139, 153, 156
failure, 108
ferroelectrics, 54
Ferromagnetic, iii, 140
fiber, 23, 36, 37, 40, 100, 101, 103, 104, 107, 114, 115, 119, 120, 124, 151, 163, 168
fibers, 14, 15, 17, 36, 41, 43, 45, 47, 65, 67, 70, 75, 100, 102, 104, 105, 107, 109, 114, 138, 162, 165, 168
fibrosis, 118
field-emission, 14, 131
filament, 21
fillers, 109
film, 4, 7, 26, 93, 108, 133, 136, 141, 161
films, 10, 80, 95, 115, 133, 150, 158
filters, 114, 115, 168
filtration, 38, 55, 114, 115, 116, 123, 168
FITC, 97
flow, 4, 19, 31, 115
flow rate, 115
fluid, 47, 137
fluorescein isothiocyanate (FITC), 97
fluorescence, 98
fluoride, 26, 114, 134
food, 116
fossil, 110
fossil fuel, 110
fossil fuels, 110
fracture, 106, 110
fractures, 109
free energy, 59
freezing, 2
friction, 109
fuel, 110, 166
fuel cell, 110, 166
fullerene, 166
funding, 117

G

GaAs, 2, 9, 12, 16, 129
gallium, 32, 37, 133, 137, 157
Gamma, 144
GaP, 9, 12, 32
gas, 4, 10, 15, 19, 21, 32, 78, 90, 93, 94, 159, 160
gas phase, 21
gas sensors, 78, 160
gases, 91, 92
gel, xvi, 36, 42, 51, 52, 53, 54, 55, 57, 59, 61, 62, 63, 64, 65, 66, 67, 68, 70, 109, 124, 138, 148, 149, 150
gelatin, 164
gels, 150
gene, 88, 95, 98, 161, 162
gene expression, 88
generation, 78, 85, 164
genes, 95
germanium, 47, 130
GFP, 96
glass, 51, 55, 168
glioma, 163
glucose, 46, 89, 159
glutathione, 89
glycol, 20, 141
glycopolymers, 168
gold, 4, 9, 10, 12, 31, 47, 76, 89, 94, 95, 96, 128, 134, 136, 137, 145, 153, 159, 160, 161, 170
gold nanoparticles, 4, 89, 137, 153
granulomas, 119
graphite, 111, 166
groups, 59, 60, 95, 97, 107, 111
growth, 1, 2, 3, 4, 5, 6, 8, 9, 10, 11, 12, 13, 14, 15, 19, 21, 26, 31, 37, 42, 45, 48, 49, 51, 55, 59, 62, 67, 101, 103, 109, 110, 123, 125, 126, 128, 129, 130, 131, 132, 133, 142, 143, 144, 153, 154, 155, 163
growth factor, 101, 163
growth factors, 101
growth mechanism, 13, 125, 128, 129, 130
growth rate, 10
growth temperature, 37
growth time, 6

H

handling, 118, 121, 171
hardness, 108
harvesting, 85, 157
hazards, 124
healing, 104
health, 117, 120, 124
health effects, 120
heat, 37, 143
heating, 4, 42, 43, 49
heavy metal, 170
hematite, 160
heterogeneous, 55, 56, 144
heterojunctions, 9, 83
hexane, 31
high temperature, 1, 2, 13, 14, 20, 21, 22, 61, 127
high-speed, 77
homogeneity, 54
host, 20, 31
HRTEM, 39
human, 96, 110, 119, 121, 166, 170, 171
humans, 119
humidity, 93, 94, 160
hybrid, 35, 43, 67, 109, 132, 138, 141, 147, 165
hybridization, 88, 89, 90
hydrate, 45, 144
hydrazine, 45
hydro, 91, 115, 151
hydrocarbon, 73, 94, 155, 159, 160
hydrocarbons, 91
hydrogels, 163
hydrogen, 4, 25, 33, 39, 45, 59, 60, 90, 92, 110, 111, 112, 159, 166, 167, 168, 169, 170
hydrogen bonds, 59
hydrogen gas, 91
hydrogen peroxide, 159
hydrogen sulfide, 33
hydrogenation, 169
hydrolysis, 51, 53

hydrolyzed, 38
hydrophilic, 115, 151
hydrophobic, 151
hydrothermal, xvi, 42, 43, 45, 46, 47, 48, 49, 51, 52, 55, 58, 59, 61, 62, 66, 67, 92, 124, 136, 139, 140, 141, 142, 143, 144, 145, 146, 148, 149, 169
hydrothermal process, 45, 46, 48, 52, 55, 92, 140, 141, 142, 144, 145
hydrothermal synthesis, 47, 55, 58, 59, 61, 140, 142, 143, 144, 148
hydroxide, 38, 43, 59, 65, 115, 136, 139, 141, 150, 168
hydroxides, 42, 43, 45, 55, 59, 62
hydroxyapatite, 47, 103, 109, 139, 145, 164, 166
hydroxyl, 59
hydroxyl groups, 59
hygiene, 121

I

images, 4, 6, 7, 8, 15, 29, 30, 35, 40, 44, 53, 57, 58, 64, 68, 69, 71, 86, 102, 106
imaging, 9, 96, 98, 99, 161
immersion, 37
immunogenicity, 96
implants, 100
implementation, 121
impregnation, 25, 31, 37, 41, 133
impurities, 2
in situ, 86, 88, 164
in transition, 150
in vitro, 96, 104, 117, 163, 166, 171
in vivo, 104, 161, 170, 171
inactive, 23, 85
incubation, 101
indication, 8
industrial, 66, 73, 117, 121, 170
industrial application, 73
inert, 14, 19, 21
inertness, 4
infancy, 117
inflammation, 118
inflammatory, 119, 120, 171
inflammatory mediators, 171
infrared, 96, 131, 161
inhalation, 120
inhibition, 118
injection, 82, 96
injury, x, 100
inoculation, 162
inorganic, 4, 43, 48, 51, 67, 70, 114, 128, 132, 141, 143, 147, 166, 167
InP, 5, 9, 12, 128
integrated circuits, 85
integration, 89, 107, 165
interaction, 27, 72, 166
interactions, 170
interface, 108, 125
interference, 137
internalization, 98
interstitials, 111
intravenous, 96
intrinsic, 12
iodine, 102, 163
ionic, 154
ions, 53, 106, 169
iron, 3, 6, 33, 97, 133, 144, 152, 161, 162, 169
irradiation, 73, 154
island, 6
island formation, 6

J

Jung, 167

K

kidney, 96, 115
kidney dialysis, 115
KOH, 27, 43, 113, 167

L

label-free, 88, 158
lamellar, 132
Langmuir, 133, 139, 144, 145, 153

laser, xvi, 9, 13, 96, 127, 149, 161
laser ablation, xvi, 9, 13, 127
lattice, 59
LED, 83
lesions, 119
life style, 121
ligand, 114
light scattering, 170
light-emitting diodes, 9, 83
limitation, 36
limitations, 113
linear, 23, 34, 72, 88, 89, 158
linkage, 95
liquid phase, 42, 45, 50
liquids, 15, 70, 151
lithium, 35, 113, 137, 159, 160, 167, 168
lithography, 76
liver, 96
low temperatures, 22, 49, 130, 155
low-temperature, 142, 147
lubricants, 108
Luciferase, 96
lumen, 119
luminescence, 153, 161
lung, 118, 119, 120, 170, 171
lung cancer, 170
lung disease, 120
lungs, 119
lysozyme, 101

M

machines, 158
macromolecules, 80
macrophages, 119
magnetic, x, 138, 154, 162
magnetic field, 154
magnetism, 160
magnetite, 97
mammalian cell, 96, 103, 170
mammalian cells, 103, 170
manganese, 45, 142
manipulation, 84, 121
manufacturing, 66, 117
MAS, 63, 65
matrix, 12, 30, 100, 101, 104, 105, 106, 108, 109, 162, 165
measurement, 84, 86, 157
measures, 121
mechanical properties, 31, 107, 109, 165, 166
media, 48, 114, 146, 167, 168
mediated gene delivery, 95
mediators, 171
MEH-PPV, 67, 69
melt, 67
melts, 151
membranes, 93, 114, 135, 136, 168
memory, 84
MEMS, iv
mercury, 94, 160
mesoporous materials, 23, 26, 132
metal nanoparticles, 27, 135, 170
metal oxide, 8, 36, 37, 43, 73, 114, 124, 138, 139, 146, 155, 159, 168, 170
metal oxide nanofibers, 43, 114, 168
metal oxides, 73, 159
metal salts, 51
metal-metal interfaces, 28
metals, 3, 12, 43
metal-semiconductor, 156
MgB_2, 51, 53, 148
micelles, 36, 45, 72, 138, 139, 142, 145
microemulsion, 73, 142, 155
microorganism, 23
microorganisms, 103, 115
microscope, 83
microscopy, 49, 76, 79
microstructure, 145, 166
microtubules, 137
microwave, 73, 154, 156, 167
migration, 21
mineralized, 105
mixing, 20, 99, 106, 165
mobility, 78, 79, 108, 157
modeling, 156
modulation, 80
modulus, 31, 106, 107
moisture, 21, 103
molar ratio, 49

molecules, 27, 59, 87, 95, 97, 100, 105, 135, 162, 163
molybdenum, 3, 144
monolayer, 80
monomer, 72
Moon, 157
morphological, 101
morphology, xvi, 3, 7, 10, 14, 20, 21, 36, 37, 39, 41, 43, 45, 50, 52, 60, 61, 65, 67, 104, 119
motion, 14
motivation, 104
multiwalled carbon nanotubes, 106, 110, 118, 137, 164, 167
muscle, 80
muscles, 108, 165
musculoskeletal, 108
musculoskeletal system, 108

N

NaCl, 20
nafion, 160
nanobelts, xvi, 2, 3, 10, 18, 20, 21, 22, 43, 44, 45, 72, 131, 132, 142, 143, 144, 145, 153, 157
nanocapsules, 97, 161, 162
nanoclusters, 5, 8, 12, 27
nanocomposites, iii, iv, vi, 103, 104, 106, 148, 151, 165
nanocrystalline, 131
nanocrystals, v, 133, 148
nanodevices, xv, 9, 36, 75, 76, 77, 80, 83, 85, 87, 123, 156, 157, 158
nanoelectronics, 83, 84
nanofabrication, 28
nanofibers, xvi, 1, 2, 3, 5, 8, 9, 12, 13, 19, 22, 23, 25, 26, 31, 32, 36, 37, 39, 41, 42, 43, 45, 46, 47, 48, 51, 52, 53, 54, 61, 66, 67, 69, 70, 71, 73, 75, 76, 80, 87, 99, 101, 102, 104, 105, 106, 107, 108, 109, 110, 111, 113, 114, 116, 118, 119, 120, 123, 126, 129, 132, 133, 135, 136, 138, 139, 140, 141, 142, 143, 144, 145, 146, 149, 150, 151, 152, 153, 154, 155, 159, 160, 162, 163, 164, 165, 166, 168, 169, 170
nanofillers, 109
nanojunctions, 83
nanomaterials, xv, 42, 66, 76, 111, 116, 118, 121, 123, 170, 171
nanometer, 8, 76
nanometers, 27, 32, 44, 67, 100, 115
nanoparticles, xv, 4, 9, 13, 23, 27, 34, 35, 50, 71, 75, 87, 89, 103, 116, 117, 123, 134, 135, 137, 140, 152, 153, 154, 158, 159, 162, 163, 164, 167, 169, 170, 171
nanoribbons, xvi, 65, 68, 147, 150
nanorods, xv, 2, 10, 21, 23, 25, 32, 33, 36, 45, 49, 50, 61, 63, 65, 76, 85, 89, 92, 94, 95, 96, 105, 109, 118, 126, 131, 133, 136, 137, 138, 140, 141, 142, 143, 144, 145, 146, 147, 148, 150, 152, 153, 154, 157, 158, 159, 160, 161, 162, 165, 169, 170
nanoscale materials, 111, 117
nanoscience, 87, 117
nanostructured materials, xv, xvi, 21, 23, 28, 66, 73, 87, 89, 94, 97, 110, 116, 118, 120, 121, 124
nanostructures, v, vi, 2, 4, 8, 23, 32, 41, 43, 49, 50, 60, 62, 71, 95, 98, 99, 105, 112, 121, 125, 130, 141, 145, 152, 163, 166
nanosystems, 76
nanotechnology, 42, 66, 73, 76, 87, 111, 116, 117, 121, 162, 171
nanotube, 26, 30, 32, 33, 39, 77, 99, 106, 107, 112, 115, 118, 133, 135, 137, 155, 156, 159, 165, 166, 168
nanotubes, xv, 2, 20, 23, 25, 27, 29, 30, 31, 33, 34, 36, 37, 39, 40, 41, 45, 49, 53, 72, 76, 79, 89, 95, 97, 98, 105, 106, 107, 110, 112, 118, 119, 120, 121, 124, 127, 133, 134, 135, 136, 137, 138, 139, 140, 141, 143, 144, 146, 149, 154, 155, 159, 162, 165, 166, 167, 168, 170, 172
nanowires, xv, 2, 3, 4, 6, 7, 8, 9, 10, 11, 12, 13, 15, 16, 20, 21, 22, 23, 25, 26, 31, 32, 34, 35, 36, 37, 38, 39, 41, 42, 45, 46, 47, 48, 49, 50, 51, 52, 53, 55, 58, 59, 60, 70,

71, 76, 77, 79, 80, 83, 84, 85, 89, 90, 91, 93, 94, 95, 118, 119, 125, 126, 127, 128, 129, 130, 131, 132, 133, 134, 135, 136, 137, 138, 139, 140, 141, 142, 143, 144, 145, 146, 147, 148, 149, 150, 153, 154, 155, 156, 157, 158, 159, 160, 161, 167, 169
natural, 105, 164
nerve, 163, 164
nerve growth factor, 163
network, 154
New York, ix, xi
nickel, 3, 95, 141
NiO, 52, 148
niobium, 3
nitrate, 37, 46, 48, 53
nitride, 25, 32, 49, 113, 133, 137, 146, 149, 153, 157, 167
nitrogen, 14, 31, 93, 146, 155, 160
nitrogen oxides, 93, 160
NMR, 63, 65, 150
noise, 156
novelty, 54
n-type, 83, 156
nuclear, 162
nucleation, 2, 13, 125, 144
nuclei, 13
nucleic acid, 88, 97
nylon, 152, 170

O

observations, 49
oil, 168
oligodeoxynucleotides, 105
oligosaccharide, 89, 159
one dimension, 155
optical, 70, 72, 75, 88, 94, 129, 138, 143, 152, 161
optical activity, 72
optical properties, 129, 138, 143, 152
optoelectronic, xv
organ, 129, 171
organic, 22, 27, 37, 43, 48, 51, 55, 59, 72, 107, 109, 115, 120, 132, 140, 141, 145, 146, 147, 170
organic compounds, 51, 170
organic solvent, 37, 48, 51, 59, 145
organic solvents, 48, 145
organometallic, 129
orientation, 70, 108, 135, 146, 150
osteoblasts, 166
Ostwald ripening, 45
oxidation, 7, 13, 39, 41, 76, 136, 139, 169, 170
oxidative, 126
oxide, xvi, 2, 3, 4, 6, 8, 13, 19, 23, 26, 31, 32, 33, 37, 39, 40, 42, 45, 53, 54, 66, 76, 101, 126, 127, 130, 131, 133, 134, 139, 141, 142, 144, 149, 151, 152, 154, 157, 158, 159, 160, 161, 162, 163
oxide clusters, 14
oxide nanoparticles, 152, 154, 170
oxide thickness, 4
oxides, 2, 12, 19, 20, 30, 37, 43, 45, 52, 54, 73, 93, 124, 159, 169
oxygen, 5, 7, 14, 21, 33, 59, 60, 78, 93, 119, 160, 171
oxygen sensors, 94, 160

P

paclitaxel, 100, 163
palladium, 3, 28, 76, 91, 135, 155, 159
Parkinson, 148
particles, 7, 12, 28, 45, 51, 61, 107, 114, 115, 117, 120, 130, 168, 170
passivation, 80
pathogenic, 120
patients, 171
patterning, 153
PbS, 133
peptide, 72, 88, 153, 163
peptides, 97, 105
periodic, 143, 158
permeability, 115
permit, 36
perovskite, 46, 52, 54

petroleum, 91, 94
pH values, 47, 62
phagocytic, 119
phagocytosis, 119, 171
pharmaceutical, 99, 116
phase transitions, 150
phosphate, 163
phosphorus, 32
photocatalysis, 169
Photocatalytic, 169
photolithography, 85
photoluminescence, 127, 129, 136, 143, 149
photon, 161
photonic, 94, 157
physical properties, 142
physics, 155
physiological, 171
piezoelectric, 83, 157
piezoelectricity, 84
planar, 60
plasma, 10, 11, 12, 22, 73, 76, 81, 112, 123, 132, 155, 167
plasmid, 105, 162
plasmids, 96
plasmons, 159
plastic, 84, 106, 157, 160
plastics, 93
platinum, 98, 134, 158, 159
play, xv, 45, 49, 52, 53
PLD, 129
PLGA, 100, 104, 164
PMMA, 107
PNA, 88
pollutant, 116
pollution, 124
poly(lactic-*co*-glycolic acid), 100
poly(methyl methacrylate), 107, 165
poly(vinyl chloride), 165
polyaniline, 66, 80, 108, 150, 155, 159, 165
polycarbonate, 26
polycrystalline, 38, 41
Polyelectrolyte, 154
polyethylene, 20
Polyethyleneglycol (PEG), 96
polyimide, 151
polymer, 31, 34, 36, 57, 67, 70, 72, 73, 80, 99, 101, 106, 107, 108, 115, 124, 136, 138, 139, 144, 151, 154, 157, 163, 168, 169
polymer chains, 108
polymer composites, 70
polymer materials, 99
polymer matrix, 108
polymer networks, 72
polymer solutions, 151
polymerization, 51
polymers, 67, 72, 80, 106, 151, 157
polypropylene, 106, 165
polystyrene, 34, 106, 114, 163, 165
pore, 23, 25, 26, 27, 31, 41, 134, 135, 144
pores, 25, 27, 170
porous, xvi, 25, 26, 28, 53, 72, 93, 97, 113, 114, 134, 135, 136, 149, 152, 168
porous materials, 25
potassium, 140
powder, 13, 15, 19, 20, 60, 65, 73, 116
powders, 5, 19, 61, 138
power, 10, 84, 85, 157
PPS, 109
precipitation, 2, 13, 62
pressure, 4, 19, 20, 42, 48, 93, 111, 112
PRI, 132
probe, 88, 98
production, 41, 42, 47, 50, 54, 55, 66, 96, 118, 121, 124, 153
pro-inflammatory, 100, 119
pro-inflammatory response, 100
proliferation, 110, 118
propane, 46
propionic acid, 95
propulsion, 109
propylene, 108
protection, 110, 115
protein, 89, 95, 97, 101, 103, 104, 105, 114, 163, 164, 168
proteins, 89, 95, 96, 97, 103, 105, 115
protocol, 146
pseudo, 73, 155

p-type, 54, 83, 149, 157
pulse, 7, 85, 133
pulsed laser, 10
pulses, 85, 92
purification, 115
PVA, 103, 151, 152
PVP, 54, 67, 69, 102
pyrolysis, 72
pyrrole, 72

Q

quantum, xv, 77, 142, 148
quantum computing, 77
quantum dot, 77
quantum phenomena, xv
quartz, 52
quasi-linear, 93

R

radiation, 103
radiofrequency, 28
Raman, 95, 147
Raman scattering, 147
random, 79, 84
random access, 84
range, 20, 26, 28, 50, 54, 75, 81, 84, 88, 89, 90, 92, 93, 109, 115, 116, 156
rare earth, 139
rat, 117
reactant, 41, 45
reactants, 48
reaction mechanism, 166
reaction medium, 47
reaction time, 47
reactive oxygen, 119, 171
reactive oxygen species (ROS), 119, 171
reagent, 23, 46, 52
real time, 88
reality, 158
receptors, 98
recognition, 114, 168
recombination, 132
recovery, 90, 93, 94
recrystallized, 72
rectification, 83
redox, 135
regeneration, 103, 105, 164
reinforcement, xv, 47, 75, 106, 107, 108, 109, 110, 123
reparation, 130, 141, 142, 146, 154, 163, 164
replication, 135
resistance, 88, 90, 92, 94
resistive, 76
resolution, 9, 71
respiratory, 102, 118, 119, 120
response time, 94
retardation, 163
Reynolds, 171
rhombohedral, 54
risk, xvi, 96, 120
risk assessment, xvi
risks, 117, 120, 162
rods, 59, 71
room temperature, 26, 38, 52, 91, 93, 99, 111, 112, 136, 150, 154, 160, 167
room-temperature, 28, 47, 94, 135, 160, 167
ROS, 119
roughness, 109
Royal Society, 99
rubber, 164
rutile, 143

S

safety, 90, 117, 120, 121, 172
salt, 20, 81
salts, 49
sample, 56, 61, 65, 68, 112
sapphire, 5, 6
saturation, 92
SBA, xvi, 25, 133, 169
SBF, 47
scaffold, 103, 105, 164
scaffolds, 75, 104, 164
scalability, 88

Scanning electron, 79
scanning electron microscopy, 6
scattering, 94, 147, 170
Schmid, 134
Schottky, 156
Schottky barrier, 156
seed, 12
seeds, 59, 66
selectivity, 33, 91, 92, 94
Self, iv, 71, 134, 141, 153, 154, 163
self-assembling, 163
self-assembly, 26, 27, 59, 62, 71, 76, 140, 146, 153, 154
self-organization, 72, 134
SEM, 4, 6, 7, 17, 18, 20, 22, 29, 30, 35, 38, 40, 44, 48, 53, 57, 58, 65, 69, 79, 81, 82, 86, 96, 102, 106
semiconductor, 2, 5, 8, 9, 12, 25, 26, 40, 50, 54, 79, 83, 129, 133, 137, 138, 146, 147, 149, 156
semiconductors, 43, 49, 51, 75, 138, 147, 148
sensing, 90, 93, 94, 123
sensitivity, 33, 87, 89, 91, 93, 94, 158, 160, 161
sensors, 78, 87, 90, 92, 94, 158, 159, 160, 161
separation, xv, 114, 123, 168
series, 12, 21, 26, 51
serum, 104, 163
serum albumin, 163
services, x
shape, 23, 37, 41, 50, 62, 84, 99, 148, 160
sight deposit, 36
sign, 171
signals, 30, 80, 84
silane, 4, 128
silica, 25, 34, 35, 36, 46, 97, 98, 113, 127, 133, 137, 138, 144, 152, 162, 167, 169, 170
silicate, 132
silicon, 2, 4, 7, 8, 10, 13, 16, 20, 21, 39, 49, 71, 76, 80, 88, 89, 90, 93, 125, 127, 128, 129, 130, 132, 141, 146, 156, 158
silk, 107, 165
silver, 3, 9, 27, 29, 46, 47, 48, 103, 134, 140, 144, 145, 152, 163, 164
simulation, 167
simulations, 79
single walled carbon nanotubes, 171
single-crystalline, 26, 47, 132, 134, 143, 144, 145, 147, 160
single-wall carbon nanotubes, 119, 162, 171
sintering, 60, 148
SiO2, 2, 13, 14, 15, 20, 37, 40, 70, 76, 117, 138, 139
sites, 63, 167
skin, 120
SKN, 145
smart materials, 158
smoothness, 108
sodium, 52, 163
solar, 116, 170
solar cell, 170
solar cells, 170
solar collection, 116
sol-gel, xvi, 36, 42, 51, 52, 53, 54, 67, 70, 124, 138, 148, 149, 150
solid phase, 1, 12, 55, 61, 73
solid state, 2, 150
solid-state, 66, 150
solubility, 12
solvent, 38, 48, 49, 66, 67, 71, 104, 106
solvents, 48, 145
solvothermal synthesis, 146, 147
sorption, 89
species, 15, 32, 62, 114, 119, 171
specific surface, 61, 111, 168
spectroscopy, 44, 140
spectrum, 43, 63, 94
speed, 77
spheres, 8
spin, 77
sputtering, 4, 28
stability, 4, 37, 60, 77, 89, 101, 103, 108, 113, 114, 152
steady state, 109
steel, 106
stiffness, 108

Stimuli, 162
storage, 75, 111, 112, 166, 167, 168
storage media, 167
strain, 95, 108
strategies, 118
strength, 21, 31, 104, 105, 107, 108, 110, 132, 136, 164, 165
stress, 95, 106
substitution, 113
substrates, 4, 21, 28, 50, 79, 157, 160
Sudan, 166
sugar, 168
sulfate, 45
sulfur, 25, 49, 146
Sun, v, 125, 137, 139, 142, 151, 153, 154, 155, 160, 161, 166, 169, 170
superconducting, 51, 137, 148
supercritical, 141
superlattices, 9
supply, 77, 111
supramolecular, 163
surface area, 72, 87, 91, 95, 97, 98, 120
surface diffusion, 132
surface energy, 14, 59
surface properties, 144
surfactant, xvi, 45, 46, 66, 134, 144, 145, 147, 148, 154
surfactants, 42, 46, 153
swelling, 90
switching, 80, 84, 94, 161
symmetry, 136
synthesis, xvi, 1, 2, 4, 8, 9, 12, 14, 16, 19, 20, 23, 24, 28, 31, 32, 33, 41, 42, 43, 45, 46, 47, 48, 49, 50, 51, 52, 53, 54, 55, 58, 59, 61, 66, 68, 70, 73, 123, 125, 126, 127, 128, 129, 132, 133, 134, 135, 136, 137, 139, 140, 141, 142, 143, 144, 145, 146, 147, 148, 149, 150, 152, 153, 154, 155, 169, 171

T

T lymphocyte, 98
T lymphocytes, 98
tar, 159
Teflon, 43, 55, 59
TEM, 7, 8, 9, 15, 22, 33, 35, 39, 40, 53, 64, 71, 81, 96, 107
temperature, 1, 2, 4, 5, 10, 11, 13, 19, 20, 21, 22, 26, 28, 38, 42, 45, 46, 47, 49, 50, 52, 54, 72, 73, 91, 93, 99, 111, 112, 126, 127, 128, 131, 135, 136, 140, 141, 142, 145, 146, 147, 148, 150, 154, 155, 159, 160, 167, 169
tensile, 104, 106, 107, 108, 164
tensile strength, 104, 106, 107, 108
terminals, 7
therapy, 95, 161, 162
thermal decomposition, 33, 42, 43
thermal equilibrium, 4
thermal evaporation, 5, 13, 18, 19, 20, 21, 22, 131, 157
thermal oxidation, 41
thermal stability, 4, 37, 60, 108, 114
thermal treatment, 14, 27, 31, 37, 45, 52, 54
thermodynamic, 12, 166
thermodynamic properties, 166
thin film, 4, 6, 80
thin films, 80
three-dimensional, 164
threshold, 78, 161
tin, 3, 6
tin oxide, 3, 6
tissue, 47, 101, 103, 105, 151, 161, 164
tissue-engineering, 104
titania, 31, 152
titanium, 3, 54, 76, 152
Titanium, 138
titanium dioxide, 152
titanium isopropoxide, 54
TNF, 118, 119
top-down, 67
toughness, 108, 109, 110
toxic, 118
toxicity, 97, 117, 118, 119, 120, 121, 170, 171
toxicological, 121
toxicology, 120, 172
trans, 108

transfection, 96
transfer, 106, 160
transferrin, 95, 161
transformation, 55, 56, 60, 62, 66, 73, 97, 161
transistor, 76, 77, 85, 155, 156, 157, 159
transistors, 77, 79, 80, 85, 156, 157
transition, 37, 51, 71, 134, 138
transition metal, 37, 138
translocation, 120
transmission, 156
transparent, 157
transport, 84, 127, 130, 139, 158
tribological, 109, 165
trout, 171
tubular, 39
tumor, 98, 100
tumor cells, 98
tungstates, 43
tungsten, 21, 140, 160
two-dimensional, 131

U

ultraviolet, 78, 136
ultraviolet light, 78
uniform, 3, 8, 9, 10, 11, 13, 20, 23, 25, 26, 36, 37, 41, 43, 45, 46, 47, 48, 52, 55, 59, 63, 70, 110, 124, 141, 144, 146
uranium, 163
urea, 53
UV light, 93
UV radiation, 103

V

vacancies, 78
vaccination, 96
vacuum, 4, 16, 84, 123
values, 47, 60, 62, 110
vapor, xvi, 1, 10, 12, 13, 19, 20, 21, 22, 23, 32, 36, 38, 55, 73, 125, 126, 127, 128, 132, 155, 160
vapor-liquid-solid, xvi, 125, 128
vapor-solid (VS), xvi
variation, 67, 141
vascular grafts, 100
vehicles, 104
velocity, 157
vinyl chloride, 106
virus, 35
viruses, 115
viscosity, 54
visible, 79
VLS, xvi, 1, 2, 3, 4, 5, 7, 8, 9, 10, 11, 12, 13, 19, 47, 49, 67, 123, 125, 126, 127, 129

W

water, 27, 37, 38, 50, 55, 59, 60, 81, 94, 115, 116, 153, 162, 168, 169
water vapor, 38, 55
water-soluble, 115, 162
wealth, 116
wear, 108, 165
web, 103, 163
Weinberg, 168
wetting, 36
wires, 8, 10, 14, 35, 37, 75, 85, 137, 154
wool, 20
workers, 27, 59, 85, 89, 95, 112, 120, 121
workplace, 120, 172
wound healing, 102

X

XRD, 19, 43, 44, 60, 61

Y

yield, 9, 31, 32, 37, 66, 88

Z

zeolites, 55
zinc, 53, 131, 143, 154

Zinc, 148
zinc oxide, 131
zirconia, 136
zirconium, 163
ZnO, vi, 2, 12, 16, 18, 20, 31, 37, 38, 45, 48, 53, 66, 77, 79, 83, 85, 86, 91, 92, 117, 126, 130, 131, 136, 137, 138, 139, 142, 143, 144, 145, 149, 150, 157, 158, 159, 160, 169
ZnO nanorods, 45, 85, 142, 143, 144, 145, 150, 157, 159, 160
ZnO nanostructures, 145